P9-CJJ-314

Praise for *How We Talk*

"If you think grammar is all about nouns, verbs, gender, and the subjunctive, N. J. Enfield's new book will transform what you think of language as being all about. At heart, language is about communicating with others in rapid-fire conversation, and linguists have found that conversation has rules just as sentence-making does. You may have heard that *mama* and *papa* are universal words—but Enfield will teach you that *huh?* is a third one. If you want to feel sophisticated just in being able to have a two-minute conversation on the phone, *How We Talk* is the book for you."

—John McWhorter, professor of linguistics, Columbia University, and author of *The Language Hoax*, *Words on the Move* and *Talking Back, Talking Black*

"N. J. Enfield is one of the most brilliant, innovative, and insightful researchers to ever work on language as a cultural construct. *How We Talk* is a superbly readable summary of his and others' work. It is a book that anyone interested in our species, communication, and the delight of learning should read. I loved every page of it."

—Daniel L. Everett, author of *Don't Sleep, There Are Snakes* and *How Language Began*

"N. J. Enfield's new book explains how everyday conversation—language we just take for granted—is all at once both ordinary and extraordinary, and how that paradox defines our very humanity. Full of examples that feel familiar, it's nonetheless a book full of surprises, written in a straightforward, friendly style distilled from long experience of making complicated things clear."

—Michael Adams, provost professor of English, Indiana University at Bloomington, and author of *In Praise of Profanity* and *Slang*

"N. J. Enfield's *How We Talk* is a delight. The book is not about the grammar, vocabulary, or usage of language, but rather about *how we collaborate with each other* in everyday conversation. Enfield's topics range from taking turns, forestalling delays, and assuring mutual understanding, to features of talk that are universal and play a role *in* the evolution of language. *Enfield* and his colleagues have investigated everyday talk in languages, both major and minor, from every corner of the world, *so* he is a true authority on these issues. Best of all, he makes these issues come alive for us readers."

—Herbert H. Clark, Albert Ray Lang Professor of Psychology Emeritus, Stanford University

How We Talk

Also by N. J. Enfield:

Distributed Agency (coeditor)

The Utility of Meaning: What Words Mean and Why

The Cambridge Handbook of Linguistic Anthropology (coeditor)

Linguistic Epidemiology: Semantics and Grammar of Language Contact in Mainland Southeast Asia

Natural Causes of Language: Frames, Biases, and Cultural Transmission

Relationship Thinking: Agency, Enchrony, and Human Sociality

The Anatomy of Meaning: Speech, Gesture, and Composite Utterances

A Grammar of Lao

Person Reference in Interaction: Linguistic, Cultural and Social Perspectives (coeditor)

Roots of Human Sociality: Culture, Cognition and Interaction (coeditor)

Ethnosyntax: Explorations in Grammar and Culture

HOW WE TALK

THE INNER WORKINGS OF CONVERSATION

N. J. ENFIELD

BASIC BOOKS
NEW YORK

Basic Books
Hachette Book Group
1290 Avenue of the Americas, New York, NY 10104
www.basicbooks.com

Printed in the United States of America

First Edition: November 2017
Published by Basic Books, an imprint of Perseus Books, LLC, a subsidiary of Hachette Book Group, Inc.
The Hachette Speakers Bureau provides a wide range of authors for speaking events. To find out more, go to www.hachettespeakersbureau.com or call (866) 376-6591.
The publisher is not responsible for websites (or their content) that are not owned by the publisher.

Print book interior design by Jack Lenzo

The Library of Congress has cataloged the hardcover edition as follows:
Names: Enfield, N. J., 1966– author.
Title: How we talk : the inner workings of conversation / N. J. Enfield.
Description: First edition. | New York : Basic Books, 2017. | Includes bibliographical references and index.
Identifiers: LCCN 2017022136 (print) | LCCN 2017040541 (ebook) | ISBN 9780465093762 (ebook) | ISBN 9780465059942 (hardcover)
Subjects: LCSH: Discourse analysis—Psychological aspects. | Conversation. | BISAC: SCIENCE / Life Sciences / Evolution. | LANGUAGE ARTS & DISCIPLINES / Linguistics / General. | LANGUAGE ARTS & DISCIPLINES / Linguistics / Pragmatics.
Classification: LCC P302.8 (ebook) | LCC P302.8 .E54 2017 (print) | DDC 401/.41—dc23
LC record available at https://lccn.loc.gov/2017022136

ISBNs: 978-0-465-05994-2 (hardcover); 978-0-465-09376-2 (e-book)

LSC-C

10 9 8 7 6 5 4 3 2 1

For SCL, with enormous appreciation

CONTENTS

1
INTRODUCTION
WHAT IS LANGUAGE LIKE?

Here are some facts about how we talk:

- The average time that people take to respond to a question is about the same time that it takes to blink the eye: 200 milliseconds.
- A "no" answer to a question will come slower than a "yes" answer, no matter which language is spoken.
- There is a standard one-second time window for responding in conversation: It helps us gauge whether a response is fast, on time, late, or unlikely to arrive at all.

- Every 84 seconds in conversation, someone will say "Huh?," "Who?," or something similar to check on what someone just said.
- One out of every 60 words we say is "um" or "uh."

I want to argue that these facts, and others like them, take us to the core of what defines our species' unique capacity for language. This claim may seem surprising, given the more bookish concerns of mainstream research on language, such as the meanings of words and the rules of grammar. But if the fine timing of answering questions or the functions of "mm-hmm" and "Huh?" seem trivial, let me borrow from Charles Darwin's remarks on the habits of earthworms: "The subject may appear an insignificant one, but we shall see that it possesses some interest."[1] Darwin is being coy. He knows the importance of his earthworm observations—worms are essential plowers of the land—and by the end of the book he does not hold back: "It may be doubted whether there are many other animals which have played so important a part in the history of the world, as have these lowly organized creatures."[2] I feel this way about the "lowly organized" elements of language that are the topic of this book: the rules we follow when taking turns in conversation, the on-the-fly ways we deal with errors and misunderstandings, and the functions of little utterances such as "uh," "mm-hmm," and "Huh?"

Researchers in disciplines from philosophy to psychology to anthropology to linguistics have long aimed to uncover the properties of the human mind that make language possible. They have focused on trying to understand how language works: what it's like, how children learn it, how it is processed in the mind. But they have had surprisingly little to say about what language is like in the back-and-forth of everyday conversation. This makes little sense, given that conversation is where language lives and breathes. Conversation is the medium in which language is most often used. When children learn their native language, they learn it in conversation. When a language is passed down through generations, it is passed down by means of conversation. Written language is many a researcher's first point of reference, but it should not be: Most languages do not have a written form at all, and in any case, written forms—from blogs to street signs to instruction manuals—are ultimately derived from the spontaneous, self-organizing system of dialogue that we call conversation.

This means that our current scientific knowledge of language, with its emphasis on decontextualized words, phrases, and sentences, is badly out of kilter. I want to show you some of what has been overlooked or set aside in the mainstream science of language. I will argue that the inner workings of conversation have their rightful place at the center of the language sciences.

It may seem strange that linguistics—the line of research responsible for understanding language—is not the source of many of the findings I will describe in this book. In its long history, linguistics has produced extensive and reliable information for many—but not all—features of language. Linguists can say a lot about things we might observe in written documents and in monologues, such as the formal structure of sentences. But for other types of information—especially those features of language that are seen only in the wilds of interaction—surprisingly little reliable information is available in linguistic reference books.[3]

In my own research on the Lao language, I often go to my bookshelf and pick out the authoritative two-volume *Lao-English Dictionary* compiled by Allen D. Kerr and published in 1972. This wonderful book has more than 1,200 pages filled with detailed entries on Lao words, including many infrequent words with meanings like "exhalation," "necessities," "collapse," and "custard apple." But there is no entry for the word "Huh?," even though this is one of the most frequent words in spoken Lao (it occurs once every six minutes in Lao conversation). This is not the author's fault. Dictionaries and grammar books on most languages tend not to record the so-called imperfections of spoken language.[4]

When I look up a word in Kerr's dictionary, it takes me a few seconds. But as Kerr explains in his preface to the work, it took him twelve years—from 1960 to 1971—to write the book. To reduce the labor of secondary researchers like me to a few seconds each time we have a question about a Lao word, Kerr invested years of his life.

Now if it happens that I reach for Kerr's dictionary but cannot find what I am after—in this case, how do speakers of Lao say "Huh?"—then I'm out of luck. I would have to go out and find a living speaker of Lao. This might be feasible if I happen to live in Fresno, California, or Sydney, Australia. I could visit a Lao restaurant and talk to the cook or the waiter. Otherwise, I'm going to have to go all the way to Laos to ask my question.

In linguistics, we rely heavily on the published findings of long-term fieldwork by predecessors. This is no different from the situation in research on any other biological phenomenon, for example the behavior of earthworms in Darwin's careful studies. Darwin not only gathered observations and reports of earthworms' behavior; he also carried out firsthand systematic experiments on his worms to find out things that could not have been known by simply looking at them. When he wondered whether worms had a sense of hearing, he did experiments on them: "They took not the least notice of the shrill notes from a metal whistle, which was repeatedly sounded near them; nor did they of the deepest and loudest tones of a bassoon."[5] His experiments continued, and he found that his earthworms were highly sensitive to vibration: "When the pots containing two worms which had remained quite indifferent to the sound of the piano, were placed on this instrument, and the note C in the bass clef was struck, both instantly retreated into their burrows. After a time they emerged, and when G above the line in the treble clef was struck they again retreated."[6]

Darwin's systematic experiments embody the controlled hypothesis testing that is standard in behavioral science. A necessary prerequisite to devising these tests was careful and long-term observation of worms in their natural environment. Darwin's book is full of reports of what he and many others had observed of worms in their natural habitat. In this way, for every species, and for every kind of behavior, a period of close observation and description must come first. So it is with the human behavior known as language.

In some lines of linguistic research, researchers are lucky that others have already gone out and done the required years of work, writing dissertations on, or even devoting careers to, questions that others may later want answers to. When we want those answers, we go to the library. But anyone hoping to find reliable data on aspects of human conversation in the library will encounter two major problems.

The first problem is that many descriptions of languages lack any information about things like turn taking, "repair,"[7] and timing in conversation. These aspects of language are often regarded as incidental to the core concerns of linguistics. The seemingly messy back-and-forth of conversation is thought to show only imperfections or perturbations of language, without intrinsic structure or merit. Here is a famous 1965 passage by

Noam Chomsky: "Linguistic theory is concerned primarily with an ideal speaker-listener, in a completely homogeneous speech-community, who knows its language perfectly and is unaffected by such grammatically irrelevant conditions as memory limitations, distractions, shifts of attention and interest, and errors (random or characteristic) in applying his knowledge of the language in actual performance."[8] This proclamation effectively ruled out the study of topics such as conversational repair in linguistics for decades, with the result that even the most accomplished linguists have little to say about how language is used in its natural habitat.

A second problem is that when information about features of conversation is actually offered by linguists, the information is notoriously unreliable. This is because linguistic researchers do not often base their research on systematic observation of firsthand recordings of free-flowing conversation. It is difficult to collect conversational data, and even when one has such data, they are difficult to study. Moreover, people often have poor intuitions about what actually happens in language. People's beliefs about language are tainted by values instilled from formal education and by social stereotypes about what is good language and what is bad. A language teacher might say that "Huh?" is not used, or should not be used, but rather one should say "Pardon?" or "Excuse me?" But these are prescriptive statements about English, not about what actually happens in language. They are about what somebody thinks should or should not happen in language. When we obtain a firsthand recording of informal conversation in the language, we hear these things within a few minutes.

The upshot is that if you want to work on aspects of language that are informal or conversational, you can't rely on dictionaries and grammar books for the data you need. To find out how people really talk, a researcher needs a special kind of direct access to language in its wild environment. The findings and insights that we will discuss in this book are possible because of the use of sound and video recordings of social interaction in everyday life. With these recordings, we can slow conversation down, look at it repeatedly, and catch every otherwise fleeting detail. Only then do we notice the defining components of language in the wild. This is one reason why "mm-hmm" and "Huh?" have not been widely studied. Another reason is that they are not the kinds of words that many linguists have recorded and studied in detail. These words tend not to occur in more formal registers of language. They rarely occur in writing. And they tend not to be taken seriously by both scholars and native speakers. Like slang words, they are often not even considered to be real language.

We need to study conversation seriously in the science of language.

An individual's ability to learn and process language is an unbeatable skill in the animal world, but it is the teamwork of dialogue that reveals the true genius of language. Even the simplest conversation is a collaborative and precision-timed achievement by the people involved. As we shall see in this book, when two people talk, they each become an interlocking piece in a single structure, driven by something that I will call the *conversation machine.*[9]

The conversation machine consists of a set of powerful social and interpretive abilities of individuals in tandem with a set of features of communicative situations—such as the unstoppable passage of time—that puts constraints on how we talk. We will look closely at how people talk, and we will see the conversation machine in operation.

Most researchers who have studied conversation have done so from outside of linguistics,[10] yet their findings suggest good answers to the deepest question that linguists have asked: What is it that humans have, and that animals lack, that explains why only our species has language? The conversation machine provides an answer to this question. The research findings reviewed in this book show that the concept of a human conversation machine defines a universal core for language, cutting across the great variety of structure in languages worldwide.

People often say that styles of social interaction differ greatly around the world, so the claim that a universal core of language is seen in conversation might seem unlikely. But I will argue that reports of radical cultural differences in how people talk have been exaggerated, at least with respect to the essential workings of conversation. We shall see that while cultural differences in conversational style can seem striking from our subjective point of view, objectively they are minimal. Differences in, say, the way in which conversational turn-taking is organized across languages are trivial in comparison to the radical differences between languages in formal structure, at every level from sound to vocabulary to grammar.

Chomsky suggested that if a Martian scientist were to study human communication, this observer would conclude that "Earthlings speak a single language."[11] I think this is the correct conclusion, but for reasons completely different than Chomsky's. His idea was that the Martian would detect underlying commonality in the structure of grammatical phrases, despite the fact that the world's languages organize their grammatical structures in a bewildering variety of ways. But it seems obvious that this bewildering variety would be more striking to our Martian observer: Languages sound (and look, in the case of sign languages) very different from place to place, with more than 6,000 distinct tongues spoken around the world. An abstract deep structure of grammar is not where our Martian would likely detect a single "Earthling language." Instead, the Martian scientist would readily observe that from Cape Horn to Siberia, from Tasmania to Tierra del Fuego, language is strikingly similar in the back-and-forth of conversation.

Our Martian would see the hallmarks of conversation in the same form everywhere: a rapid system of turn-taking in which, mostly, one person is talking at a time; an exquisite sensitivity to the passage of time in dialogue, with a universal one-second window defining subtle distinctions between being early, on time, or late to respond; and a heavy reliance on small utterances such as "mm-hmm," "um," and "Huh?" to orchestrate the proceedings. And while our Martian scientist would see these features in all human conversation, were this observer to look for these features in communication among other species, it would not find them.

Happily, we do not need to imagine what an interplanetary observer would see when looking at language in the wild. A growing number of Earthlings are studying the conversation machine in action. We now know a lot about the fine timing of behavior in conversation and about the meanings and functions of many informal words that are crucial if conversation is going to stay on course. And our knowledge of what makes conversation universal goes deeper than these surface observations, into the shared cognition that people bring to social interaction.

Language would not be what it is without our species' highly cooperative and morally grounded ways of thinking. For the conversation machine to operate, humans apply high-level interpersonal cognition: We infer others' intentions beyond the explicit meanings of their words (in ways animals can't manage), we monitor others' personal and moral commitment to the interaction and if necessary hold them to account for that commitment, and we cooperate with others by opting for the most efficient, and usually most helpful, kinds of responses. We help each other, where necessary and possible, to stay on track in conversation. This requires not only a good deal of attention and effort; it also requires social cognitive skills that are unique to our species.

The cognition that people need for language must of course be found in the head, and in that sense, cognition for language is located in individuals. But much research on how the mind works has shown that cognition is radically distributed.[12] Much of our thinking and reasoning is not done solely between our ears. When we use our brains, we often hook them up to external systems. These may be physical objects, such as pencil and paper or smartphones, that supersize our capacities for memory and reasoning. In conversation, the external systems to which we hook ourselves up are the bodies and minds of other people.

The cognition needed for language is especially attuned to what others think, feel, and mean, and it is oriented to what the members of the social unit (the "us" currently having a conversation) are collectively doing or at least trying to do. Cognition for language is intrinsically dialogic. This point is crucial to understanding the idea of a conversation machine at the heart of language.

When we talk, we do not drive the conversation machine. The conversation machine drives us.

In the chapters that follow, we will learn what humans have that makes us able to carry out the remarkable feats of everyday dialogue. We will find out what the conversation machine is, and what it does. A good place to begin is with the idea that conversation has rules, of a kind that demands a unique brand of morally grounded social cognition.

2
CONVERSATION HAS RULES

At school we learn that language has rules. There are subjects and objects, conjugations and declensions, phrases and sentences. We know thousands of words, but alone they are not enough: We have rules for taking those words and combining them into sentences. These are the rules we refer to as grammar. Most people are not able to state many of these rules explicitly, yet people everywhere subconsciously follow the rules closely when speaking, making only occasional errors.

Besides grammar, there is another dimension to the rules that guide language: the norms of conversation. So, when someone asks a question, you should answer it. If you can't answer it, you should still respond (e.g., give a reason why you can't answer). If a third person had been asked the question, you shouldn't answer for them.

We think of these not so much as rules but as simple good manners. Yet they are more than this. These are not rules for how a person should act. They are rules for how a team player should act. The rules make sense if you think of their function as regulating the flow of conversation as a kind of group activity. In conversation, everyone involved has a set of implicit rights and duties in the interaction. This is because conversation is inherently cooperative. It is a form of joint action.

Humans' capacity for cooperative joint action is one of the defining capacities of our species' form of social life. When we cooperate, we enter into a (usually unspoken) pact to join forces toward a common goal. Through this pact, we become morally accountable to that commitment and to seeing that commitment through. Joint action is not just a way of behaving; it implies a special way of thinking. The philosopher John Searle[1] imagines a scene in which a number of people are running from different directions to take cover under a shelter in the middle of a park. He suggests two scenarios in which this might happen. In the first scenario, it has just started to rain. The people are unrelated individuals running to take shelter. Each of them is motivated to run to the same place for the same reason, and while their behavior appears coordinated, they are in fact behaving independently. They are not acting as a group. In the second scenario, the people are members of an outdoor ballet troupe, and they are engaged in a public performance that calls for them to converge on this same spot at a chosen moment. The key difference between the two scenarios has to do with what the individuals think they are doing. In the first case, the appropriate thought is that "I" am running to the shelter (and, incidentally, others happen to also be doing so). In the second case, "we" are doing it. This distinction might seem academic, but it has an important consequence: It introduces a moral commitment among the people involved.

The philosopher Margaret Gilbert explored this moral consequence of joint action in one of the simplest examples she could think of: going for a walk together.[2] Her interest was in social phenomena in general, and she considered the scenario of going for a walk together to be paradigmatic of all social phenomena. Again, we can contrast two situations that look similar on the surface: Two people are walking along side by side. In one case, the two people just happen to be going in the same direction at the same time, as on any busy city street. In the other case, the two people have agreed to go together for a walk.

Gilbert points out an important difference between these two scenarios. Suppose that one of the two walkers speeds up a bit and draws ahead of the other. In the first scenario, this might not be noticed at all. But in the second scenario, the person who is walking ahead might be in for what Gilbert calls a mild rebuke: "You are going to have to slow down, I can't keep up with you!"

By definition, joint action introduces rights and duties.[3] As Gilbert says of the two people on a walk together, "each has a right to the other's attention and corrective action."[4] Each person has a moral duty to ensure that they are doing their part appropriately so that if, for example, one person draws ahead, the other may hold them to account. The duty to stay involved in joint action and the corresponding right to rebuke those who do not stay involved underlie the ground rules of language. Let us look at some examples.

Questions are a universal feature of human language. The specific grammatical rules for how questions are formed—both of the yes/no type and the who/where/what type—can vary widely from language to language. Here I am focusing not on how questions are grammatically constructed but on the ways in which questions function in social interaction. As with any joint action, questions create commitments and associated moral duties.

Suppose that I say to you "What time is it?" You are suddenly saddled with moral obligations. The first is that you can't just stay silent. Whether or not you know the answer, you should respond. Or at least, if you do not respond, you will do so with the awareness that I have a right to rebuke you, at least mildly.

In an example from a recorded telephone call, a grandmother concerned for her granddaughter's health is urging the granddaughter to go and see a doctor. The grandmother says, "I can't stand idly by and see you destroy yourself." Here is an extract from the rest of the conversation (GM = grandmother, GD = granddaughter):[5]

1. **GM:** Now will you do that for me?
2. (silence for 2.5 seconds)
3. **GM:** Honey?
4. **GD:** What.
5. **GM:** Will you do that?
6. **GD:** Well—Grandma it's gonna be so expensive to go talk to some dumb doctor.

After the grandmother asks a direct question, there is no response. But the granddaughter has a duty to respond, so the grandmother is entitled to pursue a response, which she does, until the granddaughter has fulfilled her duty. The grandmother does not explicitly rebuke her granddaughter for failing to respond, but her pursuit can be interpreted as one. When the granddaughter does respond, in line 6, she does not directly answer the question, but she does fulfill her duty as a questionee. This is similar to what happens when we are asked the time: We are not obliged to know the time, but we should state it if we do know, and if we do not know, then we should say so.

In another example,[6] Person A is within their rights to follow up on their question not once but twice before getting the answer that they required from Person B:

1. **A:** Is there something bothering you or not?
2. (1 second silence)
3. **A:** Yes or no.
4. (1.5 seconds silence)
5. **A:** Eh?
6. **B:** No.

Sometimes people are more explicit in their rebukes than the grandmother in the earlier case. Here is an example from a recording of men in a group therapy session:[7]

1. **Roger:** But tell me is everybody like that or am I just out of it?
2. **Ken:** Not to change the subject, but—
3. **Roger:** Well don't change the subject, answer me.
4. **Ken:** I'm on the subject, but I mean not to interrupt you but . . .

Roger asks a question. When Ken appears to be doing something other than give an answer, Roger launches a clear rebuke, telling him both what not to do and what he should do. Again, one person is entitled to rebuke another for failing to fulfill their part of a social pact.

Another piece of the social pact that a question creates is not concerned with the response that should come, but with the person who should supply it. Here is an example from a conversation between three young women, Amy, Ruth, and Olive. They are gossiping about a mutual acquaintance:[8]

1. **Amy (asking Olive):** Does she call you and conversate with you on your phone?
2. **Ruth:** No, that'd be wasting minutes.
3. **Amy:** I want Olive to answer the damn question, don't answer for her.
4. **Ruth:** OK I'm sorry.
5. **Olive:** She called me once to see if my mother had thrown a fit but no, other than that . . .

When Ruth jumped in and tried to answer the question that was addressed to Olive, Amy gave her more than a mild rebuke. Examples like these show that when people use questions in everyday conversation, they create moral commitments. They create rights to rebuke when people don't do what they've effectively signed up to do by virtue of the simple fact that they are engaged in a conversation. The rules become visible only when someone departs from the implicit norms.

These examples show that just as people are mostly willing to follow the rules regarding answering questions, people are also willing to enforce those rules when it seems they are being broken. A kind of social coercion operates here. It is not as when under oath in a court of law, but it seems that even in the most mundane settings of everyday talk, we still take the obligations seriously.

This doesn't mean that we can't cheat the system. As we have seen from the above cases, a person who is asked a question is obliged to provide a response, preferably an informative answer. But as any politician knows, when it comes to the *content* of an answer, there is leeway. As former US Secretary of Defense Robert McNamara put it, "Don't answer the question you were asked. Answer the question you wish you were asked."[9]

Cheating in the question-answer game is possible because we can effectively follow the rule by just giving an answer. We don't necessarily have to give what was asked for. Here is an exchange between journalist Judy Woodruff and US Senator Dan Quayle during an October 1988 vice-presidential debate:[10]

Woodruff: Your leader in the Senate, Bob Dole, said that a better-qualified person could have been chosen. Other Republicans have been far more critical in private. Why do you think that you have not made a more substantial impression on some of these people who have been able to observe you up close?

Quayle: The question goes to whether I'm qualified to be vice president, and in the case of a tragedy whether I'm qualified to be president.

> Qualifications for the office of vice president or president are not age alone. You must look at accomplishments, and you must look at experience.

Quayle is obliged to produce an answer, but beyond that he can pivot in various ways. Here he answers the question he wishes he had been asked.

Sometimes when a person wants to ask a question, they will let the listener know in advance that the question is coming. Here is an example from a radio talk show:[11]

1. **Caller:** I wanna ask you something.
2. I wrote a letter
3. **Host:** Mm-hmm
4. **Caller:** to the governor
5. **Host:** Mm-hmm
6. **Caller:** telling him what I thought about him.
7. Will I get an answer do you think?
8. **Host:** Yes.

The caller announces that they want to ask something. But they do not then immediately ask it. They first give the background to the question. The host gives feedback, showing that they are listening, until the question itself comes. Then the host answers. The example shows that by using these kinds of preliminaries, we can reserve or block out time for our own input in conversation, and ramp up the degree of commitment that we demand of others in conversation. Here, not only is the host obliged to answer the question, but prior to that they are required to sit tight and pay attention while the questioner gives the relevant background.

Sometimes this can mean sitting tight for quite a while before the question eventually comes. Again a caller to a radio show foreshadows the arrival of the question by saying "Let me ask you this" (highlighted in line 1), after which a long series of turns passes, requiring the host to do nothing more than keep prompting until line 34 (highlighted), when the question finally comes:[12]

1. **Caller:** Now listen Mr. Crandall, <u>let me ask you this.</u>
2. A cab. You're standing on a corner.
3. I heard you talking to a cab driver.
4. **Host:** Uh-huh.
5. **Caller:** Uh was it—it was a cab driver wasn't it?
6. **Host:** Yup.
7. **Caller:** Now you're standing on a corner.
8. **Host:** Mm-hmm.
9. **Caller:** I live up here in Queens.
10. **Host:** Mm-hmm.
11. **Caller:** Near Queens Boulevard.
12. **Host:** Mm-hmm.
13. **Caller:** I'm standing on the corner of Queens Boulevard and um 39th Street.
14. **Host:** Right?
15. **Caller:** Uh I—a cab comes along. And I wave my arm.

16. Okay, "I want ya, I want ya." You know.
17. **Host:** Mm-hmm.
18. **Caller:** Um I'm waving my arm now. Here in my living room.
19. **Host:** (laughs)
20. **Caller:** And uh he just goes right on by me.
21. **Host:** Mm-hmm.
22. **Caller:** And uh two three—about three blocks, beyond me
23. where—in the direction I'm going, there is a cab stand.
24. **Host:** Mm-hmm.
25. **Caller:** Uh there is a hospital, uh a block up
26. and there is a subway station, right there.
27. **Host:** Mm-hmm.
28. **Caller:** Uh now I could have walked, the three or four
29. blocks to that cab stand.
30. **Host:** Mm-hmm.
31. **Caller:** But I had come out of where I was
32. right there on the corner.
33. **Host:** Right?
34. **Caller:** Now is he not supposed to stop for me?

The same type of mechanism is used for other types of move such as asking favors. In the next example, Fred gives advance warning (in line 1) that a favor will be asked. By giving a go-ahead in line 2, Bea commits herself not only to considering granting the favor but also to paying attention during the lead-up and waiting for it to come to fruition. In this case the favor is not explicitly asked, but it is clear from what Fred says in line 5 that he wants help fixing the buttons on a blouse:[13]

1. **Fred:** Oh by the way I have a big favor to ask you.
2. **Bea:** Sure, go ahead.
3. **Fred:** Remember the blouse you made a couple of weeks ago?
4. **Bea:** Yah.
5. **Fred:** Well I want to wear it this weekend to Vegas but my mom's buttonholer is broken.
6. **Bea:** Fred I told ya when I made the blouse I'd do the buttonholes.
7. **Fred:** But I hate to impose.
8. **Bea:** No problem. We can do them Monday after work.

And in another case in which someone asks a favor, the advance warning that the favor will be asked is given in the form of a direct question: "Would you do me a favor?" When Jim gives the go-ahead, he is clearly not agreeing to do the favor, but he certainly is agreeing to act as a listener until the favor is made explicit (in line 7):[14]

1. **Bon:** Would you do me a favor?
2. **Jim:** Uh depends on the favor, go ahead.
3. **Bon:** Did yer mom tell you I called the other day?
4. **Jim:** No she didn't.

5. **Bon:** Well I called.
6. **Jim:** Uh-huh.
7. **Bon:** I was wondering if you'd let me borrow your gun.

The same mechanism for reserving or blocking out time in a conversation is used when launching narratives. In the following example, John announces that he wants to tell Edie something, which is followed by necessary preliminaries to the actual thing he wants to tell, with the punch line coming in line 11:[15]

1. **John:** I wanna tell you something.
2. You know that—when we went up to that place
3. to drive a car?
4. So I went back there
5. And do you know something
6. Listen to this Edie you guys get this.
7. Remember when we went to look at the cars?
8. **Edie:** Yeah.
9. **John:** We went to see the fella the next day
10. To drive the car
11. And he thought you were my son!
12. (laughter)

These ways of blocking out time in upcoming stretches of conversation form the basis of how we launch and tell narratives. Suppose that you are in conversation with someone and they say "Listen, something very, very cute happened last night at the warehouse."[16] With that one line, the person is proposing a pact to guidc and control the conversation for the next little while.[17] Among the rights and duties entailed by such a pact are the following:

- In the next turn you should give them a go-ahead signal (e.g., "What," "Uh-huh," "Oh yeah?," "What happened?").
- Then they should be allowed to speak uninterrupted for a while.
- They should stay on topic in producing a narrative whose climax fits the promised description: in this case, "the thing that happened at the warehouse was very, very cute."
- You should look at them and pay attention during this narrative, and you should show explicit signs of understanding as they proceed (e.g., saying "uh-huh," nodding).
- You should pay attention until they're done, without changing the subject or walking off.
- When they get to the punch line—the thing that counts as "cute"—show your understanding and appreciation by saying "Awww!" or something similar.
- You should then have the chance to launch a related story, by reference to the one that was just told.

At first glance, these rules for how to behave when someone is telling you something would seem to make sense simply in terms of being polite. We can easily find evidence of such rules in any of the many manuals on conversational etiquette. Sarah Annie Frost offers the following laws of relevance in her 1869 book *Frost's Laws and By-Laws of American Society*:

> To listen with interest and attention is as important in polite society as to converse well, and it is in the character of listener that the elegant refinement of a man accustomed to society will soonest prove itself.
>
> Strictly avoid anything approaching to absence of mind. There can be nothing more offensive than a pre-occupied vacant expression, an evident abstraction of self at the very time you are supposed to be listening attentively to all that is being said to you. Lord Chesterfield said: "When I see a man absent in mind, I choose to be absent in body." And there was really much reason in the remark.[18]

But there is more at stake than simply wanting to appear to have good character. Without the rules of listening to narrative in conversation, there would be no guarantee that these narratives would successfully get told. This matters because narratives in conversation are what enable us not only to share experience, but also to share our evaluations and appraisals of what goes on in our lives. This is a central function of the conversation machine: as a mechanism for social cohesion.

So let us consider how the aforementioned rules enable storytelling in conversation and allow it to serve its social functions.

A first problem that these rules help to solve is the need for the teller of the narrative to reserve floor time in advance, given that it takes time to put a story together. Everyday conversation operates on a turn-by-turn basis. There is no guarantee that a speaker will be able to go beyond the end of their current turn before another person comes in and moves the talk in a new direction. But when a narrative is under way, an appropriate listener will (or should!) refrain from attempting to introduce a new topic or to steer the discussion in a new direction, even when they are desperate to do so. We have all experienced the sense of being trapped as a listener of a story that seems not to end. Eventually, of course, we can ask them to get to the point: Just as a speaker gains the right to speak uninterrupted, similarly they have a duty not to go on for too long.

A second problem that the rules of narrative help solve has to do with the monitoring of both self and other that defines joint action. The fact that a listener is required to pay attention helps the teller know at each step that their story so far has been understood, and accepted by the listener up to that point.

A third problem that the rules help solve is that of interpersonal affiliation. Telling stories in conversation is not just about conveying information. Narratives don't merely describe what happened. They evaluate what happened, and they take a stance: Was it terrible, amazing, wrong, fun, or what? In turn, this stance taking gives other people the opportunity to take the same stances, thus giving people opportunities to strengthen their social bonds.

This is how language in conversation can function in ways analogous to behaviors of coalition formation in other social species. For example, capuchin monkeys build coalitions when they group together with other individuals to direct aggression at one another.[19] This is one common form of stance taking. As it happens, capuchin monkeys engage more often in pseudo-coalitions than real coalitions. Group mates get together in displaying aggression toward something harmless like an egg or a patch of dirt. The point is not what particular stance they take: aggression, interest, or whatever. What is important is that those in the group take the *same* stance. It is the same when people gossip about celebrities. Like the capuchins' harmless objects, famous people are low-cost targets and give us ready opportunities to affiliate. Everyday narratives such as a quick story about what happened on the train on the way to work similarly allow us to agree on things and to bond through doing so. So there is ample motivation to follow the rules of storytelling in conversation.

A different kind of motivation for following the rules is the desire to avoid the consequences of breaching them.

Rules of conversation become most visible when they are broken. This principle was behind a set of informal experiments carried out by students of the UCLA sociologist Harold Garfinkel in the 1960s. In one of these, "students were instructed to engage an acquaintance or a friend in an ordinary conversation and, without indicating that what the experimenter was asking was in any way unusual, to insist that the person clarify the sense of his commonplace remarks."[20] In normal conversation, people leave out a lot of details, assuming that others understand quite well what they mean. One of the rules of conversation is to respect this principle by not probing for details when they are not offered. Here is an example of what happened when students breached the norm:[21]

A: I had a flat tire.
B: What do you mean, you had a flat tire?
A: (Momentarily stunned) What do you mean "What do you mean?" A flat tire is a flat tire. That is what I meant. Nothing special. What a crazy question!

And another example, a married couple watching TV:[22]

A: I'm tired.
B: How are you tired, physically, mentally, or just bored?

A: I don't know. I guess physically, mainly.

B: You mean that your muscles ache, or your bones?

A: I guess so. Don't be so technical.

(more watching)

A: All these old movies have the same kind of old iron bedstead in them.

B: What do you mean? Do you mean all old movies, or some of them, or just the ones you have seen?

A: What's the matter with you? You know what I mean.

B: I wish you would be more specific.

A: You know what I mean! Drop dead!

Misfires in conversation are not just awkward; they can be socially costly. Even slight breaches of conversational norms can quickly lead to upset and personal conflict. To breach conventions like these with any regularity would be unsustainable. Indeed, breaking such rules too often can be taken as a sign of mental illness.[23]

In real-life conversation, we do not follow a script. Yet people not only manage to navigate conversations well; they are also highly sensitive to when things go wrong. This is because we are aware of a set of rules for conversing. Not only that, we are prone to sanction others who transgress those rules. Here we see two key elements of the conversation machine in action. First, there are rules that determine the structure of our interactions and that set the parameters and paths for driving conversation forward. Second, people apply their higher-order social cognition in closely monitoring others' behavior in conversation, exercising their moral rights and duties when transgressions occur.

These two elements of the conversation machine—the rules of engagement and the motivation to be accountable to those rules—are in constant interaction with another element: the unstoppable flow of time. The kinds of rules we have been discussing must not only define *how* we should respond but *when*. In the next two chapters, we will learn the importance of split-second timing in conversation.

3
SPLIT-SECOND TIMING

In research on conversation, the most widely cited study of all time is a sprawling paper published in 1974 in the flagship journal of the Linguistic Society of America. Awkwardly titled "A Simplest Systematics for the Organization of Turn-Taking for Conversation," it was the work of Harvey Sacks, Emanuel Schegloff, and Gail Jefferson, rogue sociologists who in the 1960s began developing a radical tradition of research on the microsociological phenomena of natural interaction. Many of the enduring principles of this approach, now known as *conversation analysis*,[1] are laid out here.

Sacks and colleagues started with basic observations about everyday conversations, such as those that any of us might have at the breakfast table, the bus stop, or the water cooler. These observations might seem unremarkable at first glance. Here are some:

- In conversation, speaker change occurs.
- Mostly, one person speaks at a time.
- Sometimes people speak in overlap, but never for long.
- Often the transition from one speaker to the next is tidy; there is no audible gap and no overlap.
- The order in which people speak is not predetermined; it varies.
- How long each person speaks is not predetermined; it varies.
- The length of the conversation is not specified in advance.
- What people say is not specified in advance.
- Sometimes it is clear who should speak next (e.g., when one is asked a question); at other times people may simply decide to be the next to speak.

Some of these claims are not likely to be disputed. Others have attracted controversy: For example, many scholars and laypeople alike question the claim that one person speaks at a time in conversation, citing cultures and contexts in which overlapping speech is common. We will look at this claim later in this chapter. For now, let us concentrate on English and the basic system that seems to operate in informal conversations (using examples mostly from telephone conversation).

Table 3.1

(Begin call)

Mathew:		Hello Redcah 5061?
Vera:	[+0.15s]	Hello Mathew is yer mum there luv
Mathew:	[+0.13s]	Uh no she's gone up to town
Vera:	[+0.24s]	Alright uh will you tell 'er Auntie Vera rang then
Mathew:	[-0.03s]	Yeah
Vera:	[+0.13s]	Okay. She's alright is she
Mathew:	[+0.10s]	Yeah
Vera:	[+0.07s]	Okay. Right. Bye bye luv
Mathew:	[+0.02s]	Ta-ta

(End call)

The simple properties of conversation listed above might strike the reader as so mundane as to be hardly worth remarking on. But that is exactly what makes them of such keen interest to scholars of conversation. From the participants' perspective, conversation seems to just play out in a completely haphazard way. But that is not what we observe. The to and fro of conversation is called turn-taking for a reason. It often plays out in pretty tidy fashion.

Table 3.1 shows an example of neat turn-taking from a quick and efficient phone call (with measurements given of the time gaps and overlaps between turns—the negative value represents an overlap).[2]

In this example, the timing of each speaker change—that is, the time it takes for the transition from one person finishing their turn to the other starting theirs—is tight, with only one- or two-tenths of a second elapsing between each turn in this sequence.

Of course, one neat example does not establish a language-wide pattern. Several large-scale studies have been undertaken to check whether the pattern of no-gap-no-overlap is the dominant one. In a 2006 study, together with psychologists J. P. de Ruiter and Holger Mitterer, I examined a large number of recordings of telephone conversations in Dutch. We measured the time that elapsed in each change from one speaker to the other—a total of more than 1,500 turn transitions in free-flowing

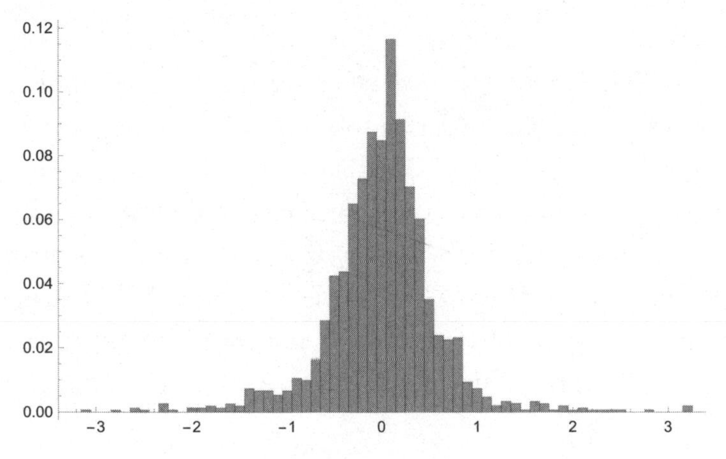

Figure 3.1 Distribution of 1,521 speaker transitions in a set of recordings of conversation in Dutch. Each bar represents a set of turn transitions within the 100-millisecond bracket indicated. The height of each bar represents the proportion of the total set of turn transitions that occurs with that timing. The highest proportion of turn transitions occurs at around a gap of 200 milliseconds. Adapted from de Ruiter, Mitterer, and Enfield 2006: 517.

conversation—and found that there was a strong spike at around the zero point, as Figure 3.1 indicates.

More than 40 percent of transitions occur within the range of a quarter-second either side of zero. These are all the cases involving both slight overlap and slight gap at a point of change between speakers. A total of 85 percent of all transitions occur within the range of three-quarters of a second either side of the zero point.

In a similar study of English, psycholinguists Stephen Levinson and Francisco Torreira measured a total of more than 20,000 transitions between speakers in conversation. They found very similar results, as displayed in Figure 3.2.[3]

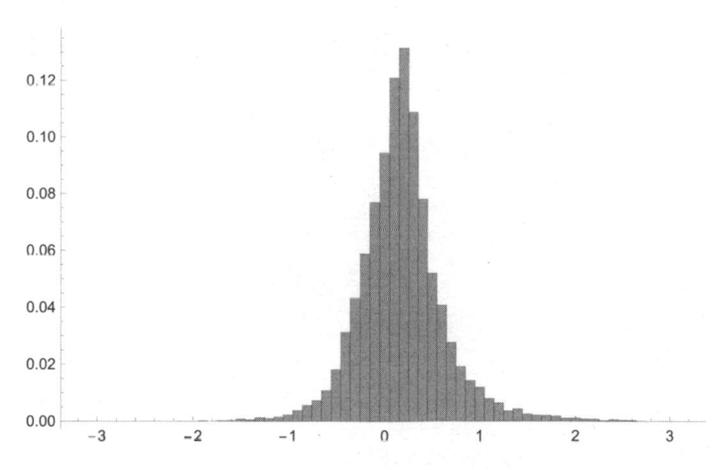

Figure 3.2 Distribution of more than 20,000 speaker transitions in a set of recordings of conversation in English. Each bar represents a set of turn transitions within the 100-millisecond bracket indicated. The height of each bar represents the proportion of the total set of turn transitions that occurs with that timing. The highest proportion of turn transitions occurs at around a gap of 200 milliseconds. Adapted from Levinson and Torreira 2015: 16.

Again, this shows that tight transitions between speakers are the norm in conversation. In this large sample of dialogue—from 38 hours of spontaneous speech, in 348 different conversations—only 3.8 percent of the time was there overlapping speech. These findings demonstrate that the one-speaker-at-a-time rule is obeyed by speakers of English and Dutch.

A further confirmation of the pattern comes from a sample of 1,500 speaker transitions in conversation in German in a 2015 study by psychologists Carina Riest, Annett Jorschick, and J. P. de Ruiter, as shown in Figure 3.3.[4]

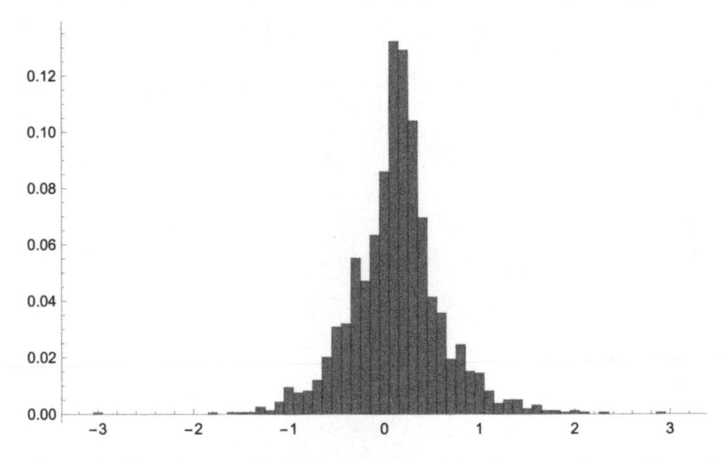

Figure 3.3 Distribution of 1,500 speaker transitions in a set of recordings of conversation in German. Each bar represents a set of turn transitions within the 100-millisecond bracket indicated. The height of each bar represents the proportion of the total set of turn transitions that occurs with that timing. The highest proportion of turn transitions occurs at around a gap of 200 milliseconds. Adapted from Riest, Jorschick, and de Ruiter 2015: 65.

These results are remarkable, for two reasons. First, they show that people generally follow the one-at-a-time rule. Most speaker transitions are timed such that there is little if any gap or overlap between speakers. Gaps and overlaps are common enough, but the vast majority of them run for less than half a second, suggesting that no-gap-no-overlap is a norm or standard. This is surprising because following the rules is not something people always do in language. Many of the rules found in grammar books—think "Don't finish a sentence with a preposition"—are broken all the time. But the one-at-a-time rule is not like this.[5] A second reason the timing results are remarkable is that they point to something psychologically very special about conversation: People are both able and willing to alternate their contributions to communication with remarkably fine timing and orderly sharing of the floor.

The studies cited above show that in large samples of conversation, people tend to transition from one speaker to the other, leaving minimal gap and minimal overlap. For people in the midst of conversation, it sounds like there is no gap. But look more closely at Figure 3.2. The highest concentration of transitions is centered not at the zero point but slightly later: at around 200 milliseconds. While there is objectively a brief silence in that gap, in fact to the human ear it sounds like no gap at all. This is not surprising: A fifth of a second is faster than the average time it takes to blink the eye.[6]

To achieve this fine timing, people must have a method for starting their turn at speaking in a conversation at exactly the right moment. According to one theory of how this is done, the person who is finishing their turn at speaking will send some kind of signal that their turn is complete, allowing the next speaker to jump in at the right moment. Researchers have proposed a number of possible signals, including a sharp drop in pitch, a gesture of the hand, or a movement of the eye gaze. But there is a problem with this signal theory. If such a signal comes at the end of a speaker's turn, there would be insufficient time for people to gear up and begin speaking as quickly as they do. This is because it takes a person well over 200 milliseconds—indeed, more than half a second—to run through all the psychological processes required to get from preparing to say something to actually articulating the words.

Decades of fine-grained research in the psychology of language production have laid bare the internal workings of how humans get "from intention to articulation," as the psychologist Willem Levelt puts it.[7] Using careful variations on psychological tests in which subjects are shown images and are required to simply name them, Levelt and colleagues have figured out the steps involved and the amount of time each step takes.[8] Suppose that a person is shown a picture of a horse. In order to name what they see, a series of things must happen. First, they need to retrieve the concept of a horse in their minds.

This takes around 175 milliseconds. Next, they need to get from that concept to the word "horse" in their mental dictionary. This takes around 75 milliseconds. Then they have to retrieve the instructions for how to pronounce this word, first by pulling up the phonological code, or the series of sounds that make up the word (taking around 80 milliseconds), then by forming these sounds into syllables (another 125 milliseconds), and finally by phonetic encoding and starting execution of the motor program for actually pronouncing the word. This adds up to a total of 600 milliseconds from the intention to speak to the sound of speaking. Now consider this in light of the fine timing of turn-taking.[9] If it takes over half a second from the launch of a piece of speech to actually speaking, and if people on average start speaking about one-fifth of a second after the prior person has finished, then it follows that intending new speakers must be gearing up to speak *well before* the other person has finished. And yet as we have seen, people are not constantly interrupting each other, nor are they leaving long gaps in the conversation. The only explanation is that next speakers somehow know in advance the moment at which the current speaker is likely to stop.

In their 1974 paper on the rules of turn-taking, Sacks and colleagues identified this ability to tell in advance when a current speaker would finish, referring to the skill as *projection*. In free-flowing conversation, people can project, or predict, in advance of the moment at which the current speaker will stop, at a point early enough that they can set their rather slow speech production mechanisms into gear so that by the time they start speaking, it is at a moment of no-gap-no-overlap. What information is available to listeners such that they are able to project the timing of speakers' turn endings? There are several possible sources of information available in speech.

A possible source of information for projecting the timing of a turn ending is the way the sentence sounds, or its prosody. Prosody captures a range of audible properties of a sentence, including pitch, intonation, stress or emphasis, speed or tempo, and pausing. These can be reduced to three elements of the vocal signal:

1. fundamental frequency, or pitch (measured in hertz)
2. duration, or length (measured in milliseconds)
3. amplitude, or loudness (measured in decibels)

The psychologist Starkey Duncan proposed in a 1972 article that various "turn-ceding cues" are used in conversation.[10] Three of these are prosodic:

1. any pitch contour other than sustained mid level (which is characteristic of speech that is continuing)
2. "drawl" on the final syllable
3. drop in pitch or loudness

When Margaret Thatcher was prime minister of the UK, her interviews with journalists had an awkwardness about them because of the fact that she was often interrupted. In a 1982 study, psychologists Geoffrey Beattie, Anne Cutler, and Mark Pearson asked why this kept happening.[11] They knew that Thatcher's speech showed an unusual pattern of intonation, so they looked more closely to see whether this played a role. They discovered that the journalists were not meaning to interrupt. Rather, when it sounded like Mrs. Thatcher was winding down to the end of her sentence, they would launch their next questions, only to find that she wasn't done. The result was that it sounded like they were interrupting her. Beattie and colleagues did a phonetic analysis of Thatcher's speech and revealed that she was actually playing by slightly different rules from her interviewers. Her intonation would feature a sharp and rapid drop toward the end of certain phrases, and this appeared to send the signal "I'm finishing my turn now," when in fact she had more to say.

Here is an example, in which the interviewer Denis Tuohy tries to begin a turn at a point when it turns out that Mrs. Thatcher was not finished:

1. **Thatcher:** If you've got the money in your pocket
   you can choose (pause) whether you spend
   it on things which attract value-added tax
   (pause) or not

2. **Tuohy:** [You s—
3. **Thatcher:** [and the main necessities don't.
4. **Tuohy:** You say a little on value-added tax . . .

The square brackets at the beginning of lines 2 and 3 in this example indicate that the two lines are spoken in overlap. In line 2, the interviewer starts speaking, but as his turn begins, Mrs. Thatcher is also speaking, continuing on from what she was saying in line 1. Tuohy quickly cuts off and yields the floor back to Mrs. Thatcher. Once she is done, in line 4, Tuohy is able to relaunch the line that he had unsuccessfully launched, and cut off, in line 2.

Beattie and colleagues noted that interruptions can occur either because one person tries to dominate the conversation or because the person has made an error in reading the turn-yielding cues. They examined the pitch contour and loudness of Mrs. Thatcher's speech in interviews. They took 40 samples, each at least a sentence long, from a recording of a lengthy interview between Tuohy and Thatcher. They played these samples to people and asked them to make a simple judgment in each case: Was her turn complete or not? Beattie and colleagues presented the extracts in various forms, including full video (sound and vision together), sound only, and written form. Here we just focus on the signals available in the sound of Thatcher's voice: the sound-only condition.

When people in the experiment heard turns that were actually complete in the interview, they mostly judged, correctly, that the turn was indeed complete. When they heard samples from the middle of Mrs. Thatcher's longer turns, nearly 70 percent of the time they accurately judged that the turns they were hearing were incomplete. With regard to both of these types of sample, something about the way the sample sounded enabled listeners to judge, more accurately than chance, whether the turn was complete, even though they had little information about the original conversational context. When they heard turns that led in the original recording to a collision of speakers—on the one hand, Thatcher continuing; on the other hand, the interviewer starting up—they judged over half of the time that Thatcher was actually finished. This supported Beattie and colleagues' hunch that the interviewer was not meaning to interrupt but had genuinely judged—like the participants in their experiment—that Mrs. Thatcher was finishing what she wanted to say.

Beattie and colleagues set out to identify just what it was about how those turns sounded that gave people the impression that Thatcher was going to finish before she actually did. They noticed that most of the utterances ended with a noticeable fall in pitch. In each recorded case, they measured two things: First, they measured how much time it took for the fall in pitch to take place; second, they measured exactly how low in pitch this fall went.

On the measure of how long it took for the fall to occur, the interrupted turns sounded very different from the turns that kept going. The interrupted turns featured a sudden drop in pitch, even more sudden than the turns that were actually complete. By contrast, the turns that kept going had a final pitch fall nearly twice as long in duration. It appears that the interviewer was using the length of Thatcher's final pitch drop as a cue that she was going to finish talking, explaining their tendency to interrupt when they did.

On the measure of how low in pitch the fall of Thatcher's turns went, the interrupted turns (which dropped to 167 hertz on average) sounded more like the turns that had not been interrupted (which dropped to 161 hertz), and unlike the turns that were actually complete in the interview (dropping to a lower point of 141 hertz).

The turns that caused turn-taking problems in Thatcher's interviews had conflicting signals about whether her turn was actually ending. The interviewer (like many of the listeners in Beattie and colleagues' experiment) was listening for the length of time taken for the drop in pitch to occur. When he heard a quicker drop in pitch at the end of the turn, he started talking, only to find that he was interrupting her. But speed of the drop in pitch was not the signal Thatcher was sending. She was signaling turn ending with a different cue, that of the pitch at which the turn would finish. Thatcher was being consistent in terms of final pitch but not in terms of speed of the drop. We could describe the resulting misfire either as a result of Thatcher sending the wrong signals or of Tuohy attending to the wrong signals. Either way, there was a persistent problem in calibration of these two people's conversational metric.

With modern methods of digital manipulation of sound recordings, this issue can be studied more closely by picking apart some of these different cues. In a study that I conducted in collaboration with psychologists J. P. de Ruiter and Holger Mitterer, we used computer manipulation to systematically remove features of the speech signal in different conditions.[12] We made recordings of informal conversations between friends (speaking Dutch), using soundproof booths. Each speaker was recorded in isolation. That way, even if a turn was interrupted, or if there was some overlap with the other speaker, the two speakers could be separated out when playing the recording back.

In our experiment, we played short extracts to people and asked them to press a button as close as possible to the moment they thought the speaker's turn would end. People are slow at pressing a button if they simply have to react to the sound of something stopping. We played sections of white noise to them and asked if they could press a button as close in time as possible to the moment that the signal ended. On average, nearly a second and a half passed between the moment that the noise stopped to the moment that their finger hit the button. But people behave very differently when they are responding to a recorded sample of speech in its natural state. On average, people in our experiment were able to hit the button within 200 milliseconds of the turn ending, around seven times faster than the white noise condition. As we have seen, 200 milliseconds is exactly the same response timing, on average, that people show when they take next turns in conversation.

We created no-pitch versions of those natural speech samples. In these versions, we used a computer program to artificially flatten the pitch of the utterance, while leaving all the other linguistic information intact—not only were all the original words clear, so were any variations in loudness and lengthening. We found that flattening the pitch of these turns made no difference to people's ability to press the button within 200 milliseconds of the turn ending. Another condition was the no-words condition. In this condition, we used a filter to mask the sound of the words so that listeners could not understand anything about the content of what was being said. But they could still hear all of the prosodic features, not only patterns of pitch, but also of loudness and lengthening—like listening to a muffled conversation through a hotel room wall. In this no-words condition, listeners performed worse at predicting the timing of the turn ending. They were twice as slow at pressing the button in anticipating the ends of turns, leaving nearly a half-second gap on average before their finger hit the button. (Note, though, that this is still much faster than the white noise condition.)

This result shows that the pitch of a turn was not a necessary or sufficient signal by itself. But it seems likely that no single signal is, all by itself, reliable for turn ending.[13]

In the pioneering paper on turn-taking by Sacks and colleagues, one of the most important cues suggested for use in predicting when somebody's turn was going to end was the grammatical structure of the turn. Look at the following utterances:

1. He's a student
2. He's a student at Radboud University
3. He's a
4. He's a student at

Taken at face value, it is clear that 1 and 2 are complete while 3 and 4 are not.

Psychologists Sara Bögels and Francisco Torreira took this observation and developed a study that aimed to see when people could be lured into thinking that a turn was ending, when in fact it was going to continue.[14] This study shows that the signals for turn ending combine several features of the sound of utterances, as well as the grammatical structure of the utterance. Bögels and Torreira's study focused on the fact that grammar alone cannot be sufficient;[15] otherwise, people would invariably think that the speaker in (2) is finished when they get to the word "student," thus leading to an interruption of the kind heard in the Thatcher interviews.

They got people to ask scripted questions in an interview situation (the study was carried out in Dutch). Some of these were short questions, such as the following:

Short question type: *So you're a student?*

The question requires a simple yes or no response. Judging by its grammatical structure, from the written form of the utterance alone, the question is complete. But if we were to hear this utterance, there is of course no sound corresponding to the question mark to tell us the question is complete.

Now consider a long question from their study:

Long question type: *So you're a student at Radboud University?*

This question is designed by the experimenters to have two distinct points at which the question might seem to be complete. The first point is where the short question ends: after the word "student." The second point is at the end, after "University." The long question type appears to be ambiguous regarding where it will end. Technically, it could be finished at "So you're a student?," yet it continues. But as Bögels and Torreira emphasize, the written form of a sentence is quite unlike the spoken form. The written form of language omits a lot of information that is present in the spoken form, in particular the prosodic features of pitch, loudness, and lengthening that Beattie and colleagues focused on in their study of Margaret Thatcher.

Bögels and Torreira measured the timing of people's responses to these two types of questions in the interviews that they recorded. In both of these questions, the same string of words occurs: *So you're a student.* They found, though, that people react differently to these same words, depending on whether they occur at the end of a turn (in the short version) or whether the speaker is going to continue (as in the long version). When people responded "yes" to the short question type, their answer was finely timed to occur close to the ending of the question, so close that the respondents must have been projecting the timing of the turn ending before it actually occurred (the placement of the response in the following examples indicates its timing relative to the question):

Scenario 1:

A. *So you're a student?* $_{S}$

B. *Yes.*

When people responded "yes" to the long question type, their answer was also finely timed to occur close to the end of the question:

Scenario 2:

A. *So you're a student at Radboud University?* $_{L}$

B. *Yes.*

In the case of the long question type, people in these conversations never misfired by saying "yes" to time with the end of the word "student." No unintended interruptions occurred, of the following kind:

Scenario 3:
A. *So you're a student at Radboud University?* $_L$
B. *Yes.*

People could plainly hear that the speaker was still going to continue after the word "student," and so Scenario 3 did not occur. Bögels and Torreira reasoned that this must be caused by a difference in how people pronounced the words *So you're a student* in the short versus the long question types. Something in the sound of the short question tells listeners that the question is going to be complete at "student." And something in the sound of the long question tells listeners that the question is *not* going to be complete at "student." To confirm this, Bögels and Torreira did a further experiment that involved cutting and splicing their natural recordings to see whether their hunch was correct. They took a recording of the long question and cut out the first part—the words *So you're a student*—then replaced it with the same words as pronounced in a recording of the short version:

[So you're a student]$_S$ *[at Radboud University]*$_L$*?*

They played two different versions of the long question—the original version and the manipulated version—and asked people to press a button when they thought the turn was going to end. When they played the long question in its original form, not a single one of the subjects (30 people) pressed the button after “student”:

Scenario 4 (= Scenario 2, above)
A. *So you’re a student at Radboud University$_L$?*
B. *0% *100%

But the spliced version had a different result:[16]

Scenario 5 (= Scenario 3, above)
A. *[So you’re a student]$_S$ [at Radboud University]$_L$?*
B. *32% *68%

A third of the respondents in Scenario 5 misfired, anticipating that the question was ending when in fact the speaker went on. This proves that there is something in the way the short version is pronounced that tells listeners that it is going to be complete at the end of the word “student.” If this information about completeness were given only by the grammatical structure of the question, then the misfires that occurred in Scenario 5 should have also occurred in Scenario 3.

Bögels and Torreira wanted to find out just what these features of pronunciation were. They measured two features of the pronunciation of the word "student," depending on whether it occurred at the end of the question, as in Scenario 1 above, or in the middle of the question, as in Scenario 2 above. They found two clear differences (shown in Figure 3.4). One was a difference in pitch. In the short questions, the word "student" was almost always pronounced at a pitch above 360 hertz, while in the long questions, the same word was almost always pronounced at a pitch below that point. Second was a difference in the length of time it took to pronounce the last syllable of "student." When the word "student" occurred at the end of the turn, as in the short questions, it was lengthened, while in the long questions it was short.

What this research shows us is that listeners in conversation are able to predict with some accuracy the future moment at which the current speaker is going to stop. With advance notice that a speaker's turn is about to end, a person who wishes to take the floor next can use the available lead time, while the current speaker's turn is still winding down, to crank up their own speech production mechanism in anticipation of speaking. This way, when they start speaking, they will do so at a point where it does not sound like they are interrupting.

The evidence that we have reviewed in support of a human system for tightly timed turn-taking in interaction has come from European languages. But people often say that conversation has a different tempo

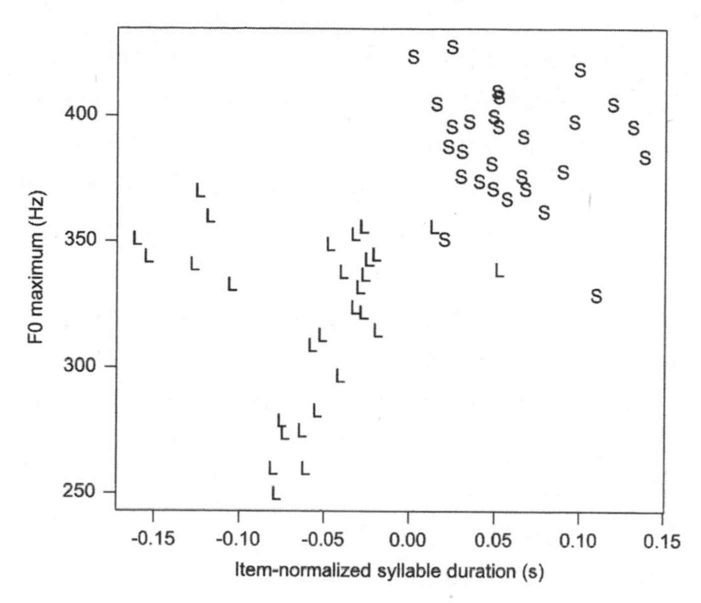

Figure 3.4 The letters "L" and "S" each represent the pitch and length of the final syllable of the word "student" in different sentences. In the short ("S") sentences, this syllable was also the last syllable of the question, while in the long ("L") sentences, this syllable was in the middle of the question. The target syllable in the short sentences was both longer in duration and higher in pitch than in the long sentences. Source: Bögels and Torreira 2015: 51.

depending on where in the world you come from. In the rural wilds of Scandinavia, the pace is said to be excruciatingly slow. There is a joke about the people of the rural southern region of Häme in Finland: "Two brothers of Häme were on their way to work in the morning. One says, 'It is here that I lost my knife.' Coming back home in the evening, the other asks, 'Your knife, did you say?'"[17] Similarly, reporting from the north of Sweden, the anthropologist Karl Reisman describes what it is like to receive visitors: "We would offer coffee. After several minutes of silence the offer would be accepted. We would tentatively ask a question. More silence, then a 'yes' or a 'no.'"[18] In other places, the pace of conversation is said to be the opposite. In the Caribbean island of Antigua, conversation is said to be anarchic, with people talking over each other.[19] And in midtown Manhattan, they say, nobody gets a word in edgewise.[20]

World travelers have strong intuitions that these stereotypes are true and that the timing of conversation is radically different around the world. There is a grain of truth in these statements, but, it turns out, it is only a small grain. Our intuitions about language are typically strong but wrong. Scientific work requires that we be skeptical of our intuitions and instead check the facts.

The fabled differences in conversational style across cultures provided me and my colleagues with a hypothesis that we could test. In a research project together with Tanya Stivers, Stephen Levinson, and a team of others,[21] we wanted to check systematically just how different the timing of conversation really is in different languages. To do this, team members traveled to field sites around the world, including sites in Italy, Namibia, Mexico, Laos, Denmark, Korea, the United States, the Netherlands, Japan, and Papua New Guinea. Each team member was an expert in the language and culture of his or her field site, and had previously spent years learning the linguistic and cultural norms of his or her distinct culture of research. The languages we examined and compared were Yélî Dnye (spoken in Papua New Guinea), Tzeltal (spoken in highland Mexico), ǂĀkhoe Haiǁom (spoken in Namibia), Lao (spoken in Laos), Korean, Japanese, Italian, Danish, Dutch, and English.

To answer the seemingly simple question of how the timing of response in conversation is organized, we had to make sure that we were measuring something comparable from culture to culture. This first meant that we had to record conversations that were similar in style. We focused exclusively on conversations in the home or the village between people who knew each other well, such as neighbors or family members. The kind of comparison we wanted to do would not work if we were using data from formal types of interaction such as village meetings, lessons, rituals, medical consultations, or legal proceedings. These more formal types of interaction are known to vary greatly depending on local cultural conventions. In a village meeting, for example, there can be special local rules about who is allowed to talk, and when. So we focused on free-flowing conversation in the home, as it seemed the most suitable kind of interaction for direct comparison.

Each researcher first collected video recordings of the kind of informal interaction we were after. Recording an hour of conversation is simple and quick. After all, it takes only an hour. The real work starts when the recording is complete. Each researcher then has to work with native speakers of each language to transcribe these interactions word by word and to faithfully translate what is being said. Rough transcription can be done quite quickly, but when the language is not your native language, and when the transcription has to be detailed, the work necessarily goes much slower. A rough rule of thumb in this work is that a minute of recorded conversation takes an hour to transcribe and translate. So while an hour of conversation takes an hour to record, it takes a minimum of sixty hours to transcribe. The background investment of time and attention from each researcher amounts to years of work before our seemingly simple research question could be checked.

When each researcher had transcribed enough conversation to find sufficient examples, we were able to define a context that would allow us to compare response time in conversation in these languages. We focused on responses to yes/no questions. These exist in all languages, and they are used often. They are simple, casual questions like "Is he a goalkeeper?," "Did he go to work?," "Is the big knife over there?," "Have you been to Birkholm?," and "Should I just print it out in advance?" The total sample of yes/no question-answer sequences in the study was around 1,500. For each case, we measured the time from the end of each question to the beginning of each answer. Figure 3.5 illustrates what we found.

The figure shows each of the ten languages plotted on a time line. The zero point of the time line, marked

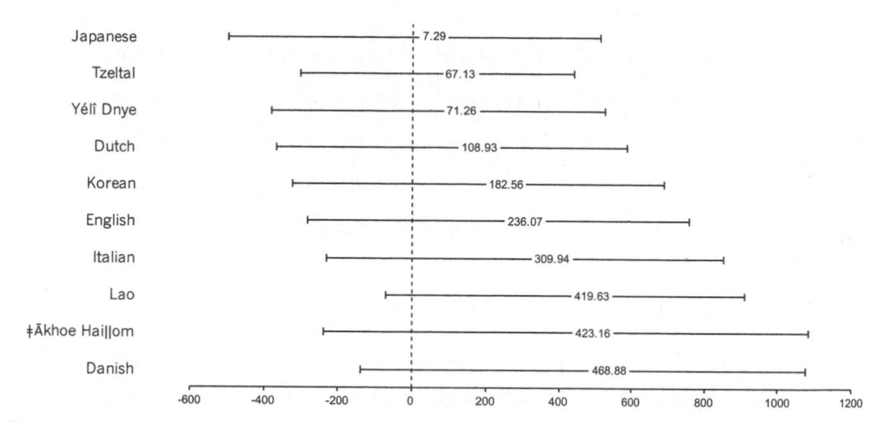

Figure 3.5 The mean time (in milliseconds) of turn transitions in conversation (±1 standard deviation) shows that speakers in all ten languages have an average offset time that is within 500 ms. However, there is a continuum of faster to slower averages across the sample. Source: Stivers et al. 2009: 10589.

by a vertical dotted line, represents the point of no-gap-no-overlap between a question and its response. For each language, we see the mean number of milliseconds that elapsed between the end of the question and the beginning of the response. English is in the middle of the field, with an average response time of 236 milliseconds. This is close to the global average for responses to yes/no questions in all languages in this study (207 milliseconds), and it is also the global average response time for speaker change in conversation. So we are seeing a very stable pattern. The figure also shows how other languages differ from English. The languages vary along a smooth cline from Japanese, at a mere 7 milliseconds average lag before a response comes, to Danish, with an average lag of close to half a second.

This might appear to lend some support to the claim, cited above, that Scandinavians are slower to respond in conversation. The Danish speakers in our study do indeed consistently time their answers to questions at a later point in time than English speakers do. But note the scale of the difference. The average Danish response is later than the average English response by less than a quarter of a second. This tells us more about people's hypersensitivity to small delays in the timing of response than it does about the supposed slowness of Danes. Actually, like speakers of all languages, Danes are very fast. In conversation, a fraction of a second from an expected point can feel like an age. We feel the quarter-second delay so much that it becomes greatly amplified in our minds. This accounts for exaggerated claims about the "slowness" of certain cultures. As we shall see in the next chapter, a half-second delay responding in English is enough to make a respondent sound distinctly hesitant.

Our cross-language study of the timing of response in conversation shows that different languages have slightly different reference points for what constitutes an on-time response. Responding on time in a Yélî Dnye conversation on Rossel Island in Papua New Guinea means getting your response out in a tenth of a second, while responding on time in a Lao conversation by the Mekong River means waiting a third of a second longer than that. Languages are calibrated differently, but when we zoom out and consider what people are doing in conversation, we see something very similar everywhere we look. In each community, the standard response centers around an agreed point that falls somewhere within a half-second window from the end of a question, with a target point varying from community to community. The human capacity to consistently land responses accurately around a common target point belies a remarkable ability, and willingness, to anticipate those points in time, using a range of cues available in the speech. And while we have seen differences between languages in how the conversation machine is calibrated, we will see in the next chapter that the most important principles governing the interpretation of how the system is used are in fact universally shared.

When we observe a universal principle in human behavior, this naturally suggests that the principle is based in the evolution of our species. It is fair to ask, then, whether other animals show comparable behavior.

Marmoset monkeys show a form of communication that researchers have referred to as turn-taking.[22] Marmosets make a long-distance contact call known as a phee call. Neuroscientist Daniel Takahashi and colleagues found that pairs of marmosets in earshot of each other would enter into a kind of turn-taking behavior: One produces a phee call, and then the other waits for them to finish before producing a subsequent phee call,

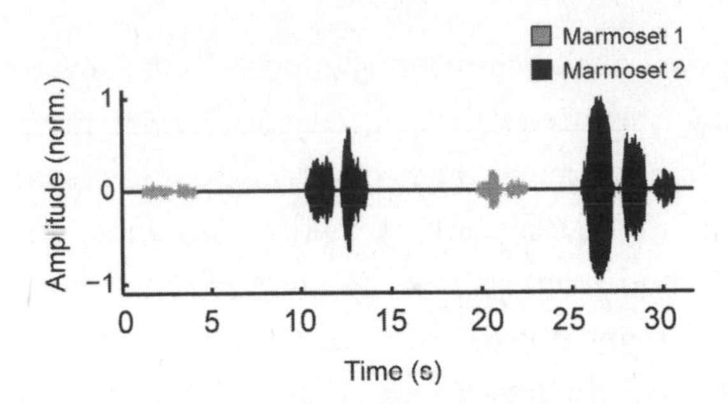

Figure 3.6 Waveform of a sequence of marmoset phee call exchanges. The *x* axis shows time (in seconds); the *y* axis shows normalized amplitude. The gray and black waveforms represent calls from Marmoset 1 and Marmoset 2, respectively. Source: Takahashi et al. 2013: 2163.

and so on, going back and forth. Figure 3.6 is an illustration of data from the experiment, showing silences of several seconds between alternating calls.

Takahashi and colleagues suggest that the marmosets' behavior resembles turn-taking in human conversation, in which people avoid overlap in speaking. But there are some big differences between the so-called turn-taking of marmosets and what humans do in conversation. One clear difference is the amount of time that elapses between "turns." For marmosets it is a consistent period of about 5–6 seconds,[23] which is about 25 times longer than the human average gap between turns. Even for marmosets, whose unit of perception might be slightly different from ours (i.e., they might experience time at a different speed),[24] this gap is truly a "waiting period," as the authors put it. One marmoset makes a phee call, the other waits in silence, and after five seconds or so they call in return. This waiting period is quite different from the tiny gap between turns in human conversational turn-taking, a gap so tight that it cannot be maintained without drawing on cognitive capacities of projection and anticipation.

The conversation machine has to ensure that transition time is minimized between turns in dialogue. But there is more to the timing of speaker transition than just aiming to start your turn at a fixed moment. The conversation machine does more than calibrate to a point. It defines a window in time, and responses can begin at various points within this window. This provides further challenges for people to time their incoming turns just right. For as we are about to learn, the window of time for transition between speakers has an internal anatomy, with subtle implications depending on whether a response sounds early, on time, or late.

4
THE ONE-SECOND WINDOW

The calibration of conversational timing and the capacity for projecting when others are going to stop speaking are crucial elements of the conversation machine. They allow people to begin talking in conversation with a minimum of delay. But just because we are capable of responding quickly doesn't explain why we so consistently *do* respond quickly. Maybe a speedy uptake of the floor is necessary if people are going to get a word in edgeways. Floor time is a finite resource. People might be starting their turn as quickly as they do simply as a way to beat out any others who might want to start talking at that moment. But often this is not a consideration. In many conversations, only two people are talking, and when a question has just been asked, there is only one person who should be speaking next. So why do people so consistently go to the effort to reply quickly?

We can find a clue towards answering this by looking at what happens when people are late to respond to questions in conversation. Here is an example:[1]

1. **A:** Do they have a good cook there?
2. (1.7 second silence)
3. **A:** Nothing special?
4. **B:** No.—Every—Everybody takes their turns.

In line 1, Person A asks a yes/no question. Instead of receiving a response within the usual average of 200 milliseconds after the question finished, there is a long silence. Person B doesn't reply. Now look at what Person A does next. Instead of repeating the question, Person A rephrases it in a quite particular way. The first version of the question was phrased somewhat neutrally. This normally creates a slight bias toward a "yes" answer. The rephrased version of the question reverses this. By putting the question in negative terms, a "no" response then becomes the natural one. This time, a response—"no"—is immediately provided.

This example shows that a delay in response, as occurs in line 2, is a signal that the respondent is sensitive to the bias of the question and is opting not to produce an answer that goes against this bias. This would account for their delay in response. In turn, the delay may indicate that Person B's answer is going to be "no." Person A's reformulation changes the bias of the question in such a way that "no" becomes the preferred response, and in this context Person B can proceed without delay.

Here is a similar case:[2]

1. **A:** What about coming here on the way?
2. (silence)
3. **A:** Or doesn't that give you enough time?
4. **B:** Well no I'm supervising here.

Person A issues a kind of request or proposal that Person B pass by and pick them up on the way. The silence in line 2 occurs when Person B should respond, but no response comes. Here we see a similar pattern as in the last example. Staying quiet, as Person B does in line 2, effectively signals that they are not going to say "yes." Accordingly, Person A anticipates this and rephrases in order to give Person B an easier way to reject the proposal. He then confirms without delay that, indeed, he doesn't have time to come by.

These examples point to an idea of *preference* in conversation, a concept introduced by early conversation analysts including the sociologists Harvey Sacks and Anita Pomerantz.[3] The absence of response in the two examples just given was, in both cases, interpreted as meaning that a dispreferred response was on the way. In the first case, the problem was saying "no" to a yes/no question rather than giving the preferred kind of answer (a "yes" answer, as in the "So you're a student?" study). In the second case, the problem was declining a request—the preferred course being to comply.

Both of the people involved in these exchanges can be seen as behaving in a prosocial way. The person who delays their response might be hoping to soften the force of their failure to comply with the other person. The other person, who, sensing this, flips their question around to facilitate the "no" answer, might be doing so to make the respondent's job easier. This shows the cooperative nature of the conversation machine at work.

I mentioned that the response was absent in line 2 of these two examples. But it might be more accurate to say that the response was delayed for too long. In both cases, the person who posed the question could have just kept waiting in silence for a response. Instead, they resumed speaking, as the floor had not actually been taken up by Person B.

Often there is a delay in responding, but it is not so long that the first person would run out of patience or assume that the second person is not going to respond at all. The sociologist Gail Jefferson looked at more than a thousand examples of delays and pauses in English conversation and noted that with few exceptions, and unless the conversation is lapsing, people are not willing to let more than about one second of silence go by. Jefferson said that one second is a "standard maximum silence" in conversation.[4]

In the ten-language study described in the last chapter, we found that in English the average time between the end of a question and the beginning of the response was less than a quarter of a second. This speedy response is around the known average time delay for speakers taking up the next turn in conversation. But there is variation within these 200-odd responses. In our comparative study, on either side of the average response time of a quarter-second was a standard deviation of plus or minus a half-second. Some of the responses came very fast, preempting the end of the question and in fact beginning around a quarter-second *before* the question ended, resulting in a brief period of overlap. Other responses were later than average, starting after a silence of around 800 milliseconds from the end of the question. To see what could account for these differences in delay time, we examined each response, sorting them into different kinds of responses that questions can get.

When one person asks a yes/no question, if the questionee responds at all, they can do one of two things. One thing is to provide an answer, that is, either a "yes" or a "no," addressing the question that was asked. If Person A asks "Was there a letter from John?," the answer might be yes or no (and these can be delivered in various forms: yes, yep, yeah, uh-huh, nods, no, nope, nah, etc.).

The other kind of response they can give is to address the question in some other relevant way, while not necessarily answering it. Typical examples convey information that explains why Person B is unable to answer the question: for instance, "I don't know" or "I haven't checked yet." When we timed these two kinds of responses—answers versus nonanswer responses—we found that the average answer (yes, no, or equivalent) started earlier than the overall average response: at around 150 milliseconds from the end of the question. The nonanswer responses started on average much later: at around 650 milliseconds after the end of the question, as shown in Figure 4.1.

There are two possible reasons why nonanswer responses start so much later. One kind of explanation has to do with cognitive processing. Person B is trying to answer the question. Suppose that it takes a person less time to decide that they know the answer than it takes for them to decide that they do not know it. The longer delay with nonanswer responses would be a necessary consequence of this. A second kind of explanation has to do with signaling. According to this explanation, if a person cannot supply the answer that the question is after, then they will intentionally delay the delivery of their response

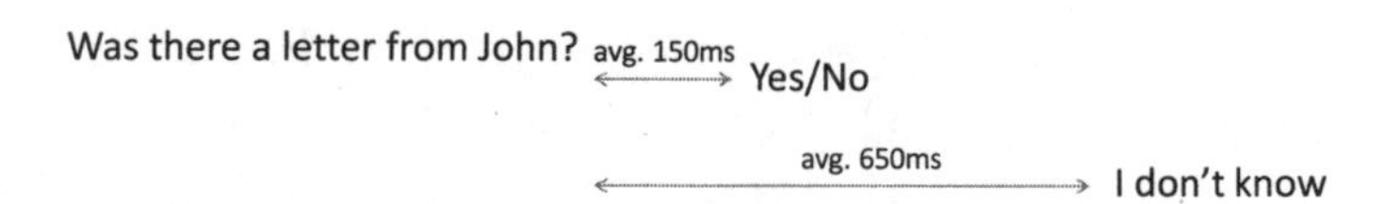

Figure 4.1 Average time delay before responses to a yes/no question across ten languages, comparing "answers" and "nonanswer responses."

as a way of sending a message that they are hesitant. This kind of signaling would be motivated in terms of the idea of conversational preference. Assuming that a person who asks a question is seeking an answer, then any response other than an answer will fail to provide what the questioner wanted. If the conversation machine urges us to be cooperative in social interaction, then we should want to avoid nonanswer responses. Just as in the "good cook" and "coming here on the way" examples above, people insert a delay prior to starting responses that seem to go against the bias of the question. But unlike in those examples above, the delay does not exceed the "standard maximum silence" of one second. The questionee does their duty in responding within that one-second time window, but they leave enough of a gap to show that what is coming is not the preferred kind of response.

When we zoom in on the responses that provided actual answers, we again find a timing difference between two types. We took all the answers and sorted them into the "yes" and "no" cases. "Yes" answers had an average delay of just 35 milliseconds, while "no" answers, while still very fast, took almost twice as long to arrive (see Figure 4.2).

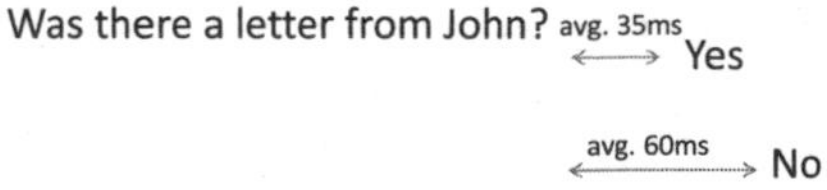

Was there a letter from John? avg. 35ms Yes

avg. 60ms No

Figure 4.2 Average time delay before responses to a yes/no question across ten languages, comparing "positive answers" and "negative answers."

Again, there are two possible explanations. A cognitive processing explanation would say that it just takes less time to decide that the answer is yes than it takes to decide that the answer is no. Call it the fast-yes effect. There is possible evidence for this in experimental psychophysics,[5] in the so-called fast-same effect. In a simple experiment, people see two fields of color, side by side. They have to press one of two buttons as fast as they can: one button if the colors are the same, the other button if they are different. It turns out that when the colors are the same, people are faster at responding accurately. For the simple dimension of color, "same" judgments are delivered within an average of 400 milliseconds of seeing the stimulus, while "different" judgments take up to 100 milliseconds longer than that.[6] If there is a link between the positively framed concepts "same" and "yes," then this effect might explain, in terms of cognitive processing, why people answer questions in conversation more quickly when the answer is "yes."

A signaling explanation of the fast-yes effect would point to a different mechanism. This explanation appeals to a general assumption that people like "yes" answers more than they like "no" answers. On this account, if a person is going to answer "no," it is a good idea to delay it a bit, for the same reason we saw a delay in the "good cook" example. "No" is likely not the answer the questioner wants to hear.

In a study of polar questions in American English, the sociologist Tanya Stivers found that of all yes/no questions that received an answer, nearly three-quarters of these were answered with "yes" (or equivalent, such as "yep," "uh-huh," head nod, etc).[7] This is a curious statistic, given that one might expect "yes" and "no" to occur in 50/50 distribution. There are two possible explanations, both based on the idea that people prefer to say "yes."

The first explanation for why "yes" answers are more common than "no" answers is that people like saying "yes" so much that they say it some of the time even when they want to say "no." Even if people were that wishy-washy, it seems doubtful that this effect would be strong enough to explain the fact that "yes" outnumbers "no" by three to one. A better explanation appeals to the *questioner's* behavior. If a questioner anticipates that people don't like saying "no," then they will try to avoid asking questions that are likely to be answered with "no." People have the choice to phrase questions just so they may be answered with "yes." Suppose that Person A asks Person B "Was the movie good?" This seems to be a neutral question, but in fact it biases a "yes" answer. If Person B arrives home from the movie with a look of annoyance on their face, Person A might well phrase the question differently. "Did the movie suck?" will also bias a "yes" answer (or at least a confirmation; the "good cook" example above shows just such a rephrasing, where "no" actually confirms what is being asked).

We have looked at two kinds of explanation for timing differences in response in conversation. Cognitive processing causes a delay simply by virtue of the fact that some tasks take longer to perform. A delay caused by processing does not necessarily indicate any form of reluctance. Consider the following example:[8]

Q: *In which sport is the Stanley Cup awarded?*
(1.4 second silence)
A: *um* (1 second silence) *hockey*

It seems pretty clear here that the answer comes late because the person is simply having a hard time remembering which sport is being referred to. We would be unlikely to infer that the person is intentionally delaying their response in order to send a signal of reluctance to give this particular answer.

Still, a timing delay caused by a delay in cognitive processing can be said to have meaning. It has meaning in the same way that a dog's baring its teeth means that it may bite. Baring the teeth is a necessary prerequisite to biting, and so when we see bared teeth this can be taken to indicate an imminent action of biting. In his 1872 book *The Expression of the Emotions in Man and Animals,* Charles Darwin described how this kind of meaning can become ritualized in animal behavior. The action of baring the teeth may have once simply been associated with the imminent act of biting (as darkness of clouds may be associated with imminent rain). But then the behavior becomes ritualized. Even in the absence of a true intention to bite, an animal can learn to bare its teeth as a way of intentionally signaling that it is in an aggressive frame of mind. Other animals can learn to interpret bared teeth this way.

Delaying a response to a question works in a similar way. Sometimes, a delay is directly caused by a straightforward problem in processing a response. In the "Stanley Cup" example just given, the delay tells us that the person is having trouble remembering. Once that association is in place, then people can intentionally manipulate the timing of their responses to send the signal "I am having trouble saying what I want to say," even when they are not literally having trouble saying it. Just as with the baring of teeth, once ritualized, this signal can be de-linked from its original underlying cause. One doesn't need to be actually having production problems to use response delay in this strategic way. Through a process of ritualization, delay of response becomes a subtle signal of reluctance, which we are able to finely control.

The interplay between the processing and the signaling explanations of delay in response is well demonstrated in a recent study by linguists Felicia Roberts and Alexander Francis.[9] In an experiment, people listened to short conversations that ended in a simple request, such as "Could you give me a ride over there?" The person replies with "Sure," but the experimenters varied the amount of time that passed before the word "Sure" was heard. People who listened were asked to give a judgment of how willing the respondent was to grant this request. When the word "Sure" occurred at around the point of average turn transition in conversation—around 200 milliseconds—people gave high judgments to the willingness of Person B to grant the request.

But when the response of "Sure" occurred after a one-second delay, people judged the person to be less willing. This finding should not be surprising to the average person, as we are all sensitive to the meaning of a delay in granting a request. But a particular finding from Roberts and Francis's study reveals something striking about the meaning of delay in response. They found that if a response of "Sure" was delayed anywhere between 100 and 500 milliseconds from the end of the request, people didn't really change their interpretation of how willing the person was to fulfill that request. But if the same response—the word "Sure," pronounced exactly the same way—was delayed by more than a half-second, people judged that the respondent seemed less willing to fulfill the request.

This reveals that the one-second window following one person's turn does not feature a smooth gradient with a direct correlation between length of delay and degree of unwillingness. Instead, people react differently to delays in different regions of the one-second window. Roberts and Francis show that people read less meaning into variations in timing in the first part of that window than they do in the second part of that window (see Figure 4.3). A difference of 300 milliseconds is not likely to be

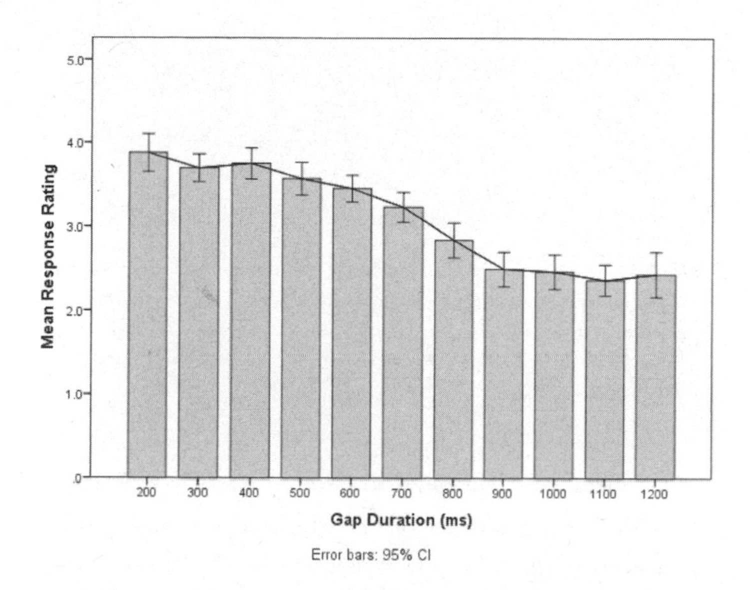

Figure 4.3 Mean listener ratings of "willingness" to comply with a request, according to duration of silent gap between requests and affirmative responses. Judgment scale ranged from 1, "not willing," to 6, "very willing." The dark line links mean values and represents slope of the decision curve. Error bars indicate standard error of the mean. Source: Roberts and Francis 2013: 475.

interpreted as a social signal if it occurs within the first half-second of response time. But if that same time lag occurs a bit later, making the difference between a delay of 500 and 800 milliseconds, then this will make a difference for its interpretation in conversation (see Figure 4.4). The 600 millisecond mark is, as Roberts and Francis put it, a "point at which social attributions emerge."[10]

The one-second window has its own anatomy. The first and second halves of the one-second window are shorter zones of time with distinct properties. We can

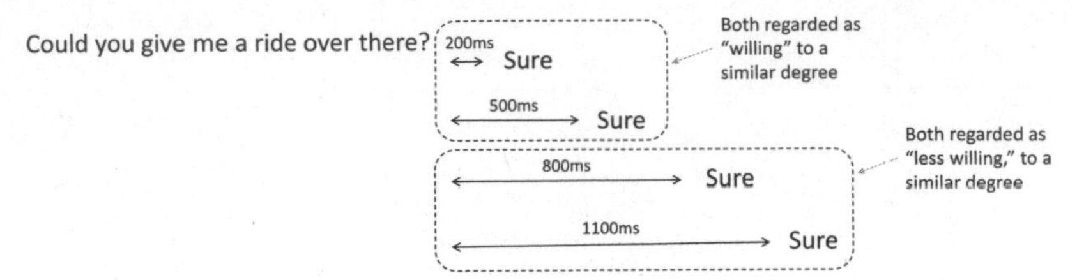

Figure 4.4 Comparison of a 300-millisecond time delay in a response that grants a request, depending on where in the one-second window it occurs: If the delay bridges the divide between the on-time zone and the late zone, it has a stronger effect of reducing the perceived willingness of the response. After Roberts and Francis 2013.

call the first half the on-time zone.[II] As we have seen, this is the zone within which the majority of speaker transitions take place in conversation. If transitions occur in this first half of the one-second window, they will sound smooth. The second half of the one-second window we can call the late zone. If transitions occur here, they tend to sound late, with associated implications of unwillingness or dispreference.

The reason that people are differently sensitive to delays in the first part versus the second part of the one-second window has to do with the relationship between processing delays and signaling delays. When people are in conversation, they are processing a lot of information, engaging in reasoning, and planning and executing speech and gesture movements, all under time pressure. All of this processing runs fast, but it still takes time. This puts more pressure on the first half of the one-second window. In the on-time zone (or transition space) that closely follows the end of a person's turn at talk, respondents will have less flexibility or control over their response timing. This is because, for example, they may not have yet fully resolved a range of processing tasks in formulating their response (retrieving words from their mental vocabulary, etc.). Similarly, they may be under the influence of various automatic processes (e.g., matching of others' bodily rhythms, or similar psychological processes having to do with prediction and timing). If this is the case, then timing perturbations in the first part of the one-second window are less likely to have been produced by intention, or to otherwise have been under the person's control. They are more likely to have been symptoms that happen to *reveal* information about the speaker rather than signals that are *intended* to convey that information.

By contrast, the second part of the one-second window is a zone in which people will have already resolved many of the things they need to process, and will have freed up resources for greater control over their own production of speech. So, if there is a small time lag within that late zone, then it is more likely that this is because the respondent wanted the time lag to be there. When a person acts with more control, we naturally read more intention into their actions. As the one-second window wears on, processing issues are less likely to be the cause of delays, so any delays can be more readily understood as intentional signals by someone who is about to speak.

This explains why Roberts and Francis found that a delay that extends into the latter part of the one-second window is more likely to be a sign of unwillingness.

When a person is going to produce a dispreferred response such as a "no" answer or a rejection of an invitation or a request, they will often delay the core content of their response not with silence, as in the Roberts and Francis experiment, but also with various fillers that delay the onset of the turn's real content. Here is an example from the closing moments of a phone call:[12]

A: If you'd care to come over and visit a little while this morning I'll give you a cup of coffee.

B: (Coughs) (breathes in) (pause) Well, that's awfully sweet of you I don't think I can make it this morning (breathes in) um (pause) I'm running an ad in the paper and and uh I have to stay near the phone.

Person A's turn is a somewhat warm invitation. Person B is going to say "no," but rather than just coming out with it, and rather than leaving an extended silence, she uses other delaying and softening tactics: first a cough, then a little in-breath, a short pause, the little word "well," some passing praise for the gesture ("That's awfully sweet of you"), and finally the declination, in hedged terms: "I don't think I can make it." Then, in addition, she supplies a reason for this declination, thereby accounting for her failure to accept the invitation. This example shows the hallmarks of putting off a dispreferred response.

In a study of the timing of dispreferred responses, linguists Kobin Kendrick and Francisco Torreira took a set of nearly 200 requests, offers, proposals, invitations, and suggestions from telephone conversations in English. They grouped the examples into two sets. In one set, the responses were of the preferred type; in the other, they were of the dispreferred type. So, for example, in relation to a proposal or suggestion—such as "Couldn't you advertise among teachers a bit?"—a preferred response would be immediate agreement, while a dispreferred response would be rejection or, as shown in Figure 4.5, a half-hearted or noncommittal agreement.

In the case of this dispreferred response, it is not merely silence that intervenes before the response proper. Kendrick and Torreira identify distinct elements that can conspire to distance the response from the thing it is responding to, with reference to four defined points, shown in Figure 4.5 with vertical lines: first, the end of the first person's turn; second, the first audible part of the response, including in-breaths; third, the first linguistic part of the response, including interjections or

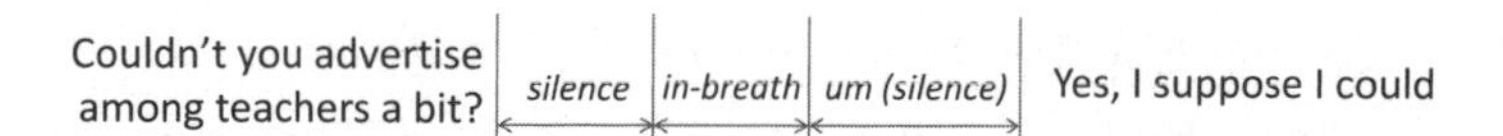

Figure 4.5 Some of the things that intervene between a question and a "dispreferred" response. Adapted from Kendrick and Torreira 2015: 12.

"particles" such as "um" or "well";[13] and fourth, the beginning of the response proper. By locating each of these points in time, Kendrick and Torreira were able to refine our understanding of how responses are delayed, with reference to three distinct measures of delay, all taken from the end of the first person's turn: first, the time of silence before the first audible part of the response; second, the time before the first particle of response; and third, the time before the first word of response that has relevant content.

When they measured how long the pure silence was before the response began, there was a roughly equal occurrence of preferred and dispreferred response types within the on-time zone, the first half of the one-second window. (In the late zone, after 750 milliseconds, dispreferred responses were more frequent.) They found that the average silence before the first audible part of a dispreferred response is in fact slightly *shorter* than in the case of a preferred response. But this is important: In nearly half of the dispreferred responses, the first sound one hears is not a word at all, but an in-breath (or click, that is, a "tut" or "tsk" sound). By contrast, preferred actions mostly do not start with these nonword sounds (only 17 percent do). So the reason why dispreferred responses average slightly shorter silences is that people are starting quicker with things that are not words (but which signal that a dispreferred response is forthcoming).

Because many dispreferred responses begin with an in-breath or click, the first word or particle comes, on average, later than the first word or particle of a preferred response. Again, the proportion of dispreferred responses is higher after about 700 milliseconds, but dispreferred responses start earlier than preferred responses on average. This turns out to be because dispreferred responses are often prefaced with interjections like "um" and "well"—things that do not yet convey anything specific about the content of the response.

When the two types of response—preferred versus dispreferred—are compared based on the measure of time that elapses before the real content of the response actually starts (i.e., minus all in-breaths and particles like "um" and "well"), then we see a real difference in the timings of preferred and dispreferred responses, in the expected direction. Half of all dispreferred responses occur in the late zone, after 500 milliseconds. Only 20 percent of the preferred responses do. Even though most responses occur well within the one-second window, the delivery of the two types is different. For dispreferred responses, they are prefaced with buffers like "um," and in this way any bad news is effectively delayed. The largest count of dispreferred responses is at 600 milliseconds, in the late zone and at precisely the "key moment in the projection of social attributions" identified by Roberts and Francis.

In the context of this basic form of signaling—dispreferred actions come a bit slower—people show a capacity for manipulating this system for special effect. Kendrick and Torreira showed that while normal acceptances come fast and normal rejections are delayed, people can exploit these expectations in conveying conversational nuances:[14]

A: Would Sunday be all right?
(approximately 600 ms silence)
B: Uh, yes as far as I know?

While this is, technically, an acceptance of the proposal of Sunday, it has the features of a dispreferred response. There is a pause of over half a second, then "uh," and then finally the response "yes." These features fit with the fact that the acceptance is qualified, as is also apparent from the addition of "as far as I know," a sort of potential disclaimer. So this is not a straight acceptance but a qualified one, and it is packaged as such.

The other type of possible special signal is a rejection that is delivered in the way an acceptance normally would be, that is, quickly and without any buffer material. First, look at how a normal acceptance is delivered:[15]

A: So uh we wondered if perhaps we'd give that a try what do you think

(100 ms silence)

B: What a good idea.

In the following case, the timing of response is virtually the same, except this time it is a rejection:[16]

A: Yeah the carnival tonight.

B: Yes.

A: Do you want to go?

(100 ms silence)

B: Oh no (300 ms pause) I'm too tired Mark.

Part of what makes this a flat rejection rather than a normal rejection is the fact that the negative part of the response comes so quickly, very early in the on-time zone of the one-second window. It is a distinct contrast from the usual pattern by which rejections are noticeably delayed. By exploiting this contrast, a person makes it clear that they are not sugarcoating the response.

In English, yes/no questions have some clear characteristics. They get answers with less delay than they get other kinds of response. Answers tend to be delivered in the on-time zone of the one-second window. Non-answers are delayed until the late zone of the one-second window, arriving on average after a delay of around 650 milliseconds. Because our cross-cultural turn-taking study was drawn from a larger comparative project on how people respond to questions, we can compare these English findings with those from other languages and see if this pattern is a quirk of English or something more general.

When all the responses to questions in our sample of conversations from the ten languages mentioned above were grouped into answers versus other things (e.g., "I don't know" or "Ask your dad"), a clear pattern emerged: Answers come faster than things that aren't answers. This pattern was shown clearly in all ten languages, as can be seen in Figure 4.6.[17]

The figure shows that in all ten languages examined, when we divide all responses to yes/no questions into two groups—answers (yes, no, or equivalents) and

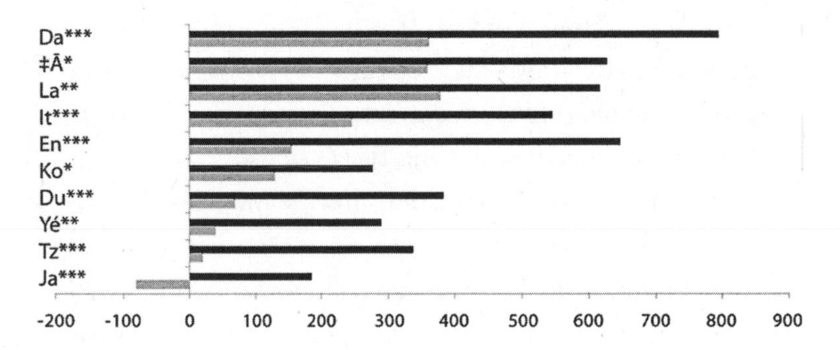

Figure 4.6 The mean time of turn transition for answers (in gray) versus nonanswer responses (in black) in ten languages. Speakers of all languages produced answers faster, on average, than they produced nonanswer responses. Milliseconds are shown on the *x* axis. Languages are arrayed along the *y* axis: Da, Danish; ‡Ā, ‡Ākhoe Hailom; La, Lao; It, Italian; En, English; Ko, Korean; Du, Dutch; Yé, Yélî-Dnye; Tz, Tzeltal; Ja, Japanese. Source: Stivers et al. 2009: 10589.

nonanswers (any other relevant response)—the answers are provided with significantly less delay. In Tzeltal, for example, an indigenous Mayan language of upland Central Mexico, answers come on average very quickly, with almost zero delay from the end of the question, while nonanswers are delayed on average for about 350 milliseconds. In ǂĀkhoe Haiǁom, an indigenous language of Namibia, answers arrive on average after a delay of around 350 milliseconds, while nonanswers arrive about 300 milliseconds later than that.

The same pattern is repeated in language after language. There is less delay before answers than before nonanswers. It is important to note, though, that the principle operates in relative terms. Just as the languages' main reference points for speaker transition in conversation are different, so too the absolute timings of answers and nonanswers are different. In the examples just noted above, a delayed response in Tzeltal (the average for nonanswers) comes at around the same speed as an on-time response in ǂĀkhoe Haiǁom (the average for answers = 350 milliseconds). But in relation to the language's locally calibrated reference point, the exact same delay principle holds.

Linguist Felicia Roberts and colleagues did a comparative study to test their earlier finding that delay in response in English is interpreted as unwillingness.[18] They ran the same experiment as before—in which they manipulated the time delay before a response of "Sure" to a request—this time in three languages: US English, Japanese, and Italian. They found that all three of these language groups had the same general response to a marked delay. If a request is granted with "Sure," manipulating the timing of that response will lead to the same interpretation across languages. A fast response will be regarded as more willing and a delayed one as less willing.

Although the lateness effect operates independently from the language being spoken or the culture of the speakers, Roberts and colleagues noted that the significance of delays can be subtly different in different cultures. They compared Italian speakers and Japanese speakers, and found that Italian speakers are slightly more sensitive to delays in the early zone of the one-second window, while Japanese speakers react slightly more to delays in the late zone.[19]

The finer details of how people time their turns in conversation reveal one of the core contributions of the conversation machine: It defines a one-second window and gives it an anatomy, dividing it into zones with specific meanings. These meanings are derived naturally from psychological by-products or symptoms of language processing, and people are now able to manipulate timing to send social signals about how a response is being packaged. In looking at examples of how people deliver dispreferred responses, we not only learn about the split-second timing of response in conversation, but we also see that various vocal sounds and interjections like "well" and "um" also play a role in packaging and postponing certain kinds of response in conversation. A close look at conversation reveals that these little words are not just throwaway tics or fillers. As we are about to see, these are subtle signals with meanings that regulate the traffic of conversation.

5
TRAFFIC SIGNALS

In any instance of communication, there is an act of production by one party (the one who sends the message) and an act of perception or comprehension by another party (the one or ones who receive the message). The conversation machine operates by linking these two acts directly. In written language, the reader does not directly witness the act of production. If, as I am writing this, I make an error or change my mind about what I want to say or how I want to say it, I can go back and edit before finalizing. You do not see which words or lines I have re-done. When we read a written text, we are usually spared the twists and turns of language production.

Conversation is different. In conversation, the speaker chooses words and puts together sentences on the fly, in real time. There are inevitable problems with things such as choice of words, pronunciation, and the content of what a speaker wants to say. And there is no script for the conversation. We do not know who will say what and when. This demands that we make constant use of little signals that regulate the traffic of talk.

Psychologists Herbert Clark and Jean Fox Tree make this point with the following example. Here is a cleaned-up, written version of a sentence from the London-Lund Corpus of Spoken English:[1]

> *Well, Mallet said he felt it would be a good thing if Oscar went.*

And here is how the original sentence was actually produced:

> *Well, (pause) I mean this (pause) uh, Mallet said Mallet was uh said something about uh you know he felt it would be a good thing if uh (pause) if Oscar went.*

Clark and Fox Tree explain the twists and turns that account for the speaker's "performance additions": The speaker "first took one direction ('Mallett said something about') and then another ('he felt it . . . '). He replaced phrases ('Mallett said' by 'Mallett was'), made clarifications (marked by 'I mean' and 'you know'), repeated words ('if if'), and added delays (silences and 'uh')."[2]

Noam Chomsky represents a widely held view in linguistics that these imperfections and signs of turbulence in spontaneous performance have little to do with language and can thus be ignored. Some researchers find them interesting to study insofar as they reveal things about how people process language. Psychologists of language have taken this view. A third view is that, as Clark and Fox Tree put it, such performance additions are "genuine parts of language."[3] They are used for regulating the traffic of conversation, telling us when to wait, why, and for how long. Clark and Fox Tree choose a good place to start in investigating this issue. They examined one of the simplest, most frequent, and perhaps most reviled little signals: "um" and its sibling, "uh."

In pioneering research on the psychological processes involved in how people produce speech in real time, psychologist Willem Levelt conducted experiments in which people were asked to look at simple maps, consisting of different colored dots connected by lines, and to verbally describe routes through these maps. To describe these routes, people constantly had to label the colors of the dots. This often led to errors of word choice. They might refer to a dot as "brown" when it was in fact pink. About half of the time, speakers do not even catch the error, or at least they don't bother to fix it. But when they do correct errors, they systematically use expressions like "um" and "uh." Here are two examples:[4]

First a bro—uh a yellow and a green disk
And from green left to pink; uh from blue left to pink

In the first case, the speaker is going to say "brown" but cuts off and replaces with "yellow." As they are gearing up to do the replacement, they start with "uh." In the second case, the speaker has produced the complete expression "from green left to pink," only then to interject "uh" and re-do the phrase replacing "green" with "blue." What is this "uh" doing here?

Clearly, "um" and "uh" are associated with problems in speaking. One view of their function is that they are simply a kind of spasm or involuntary symptom of an internal problem with speech production. Levelt takes this view, calling them "symptoms." A different view is that "um" and "uh" have an intentional message behind them. Sociologist Erving Goffman suggested that people use these words to signal to others that they are momentarily busy formulating their thoughts.[5] The motivation for giving such a signal is to convey to a listener that despite one's pause or lack of fluency, the problem is about to be corrected.

"Um" and "uh" occur in similar situations of disfluency or problems in speaking. But Clark and Fox Tree found an important difference between the two forms. In a study of the London-Lund Corpus of Spoken English, they isolated nearly 4,000 examples of "um" and "uh," and measured the time that elapsed before fluent

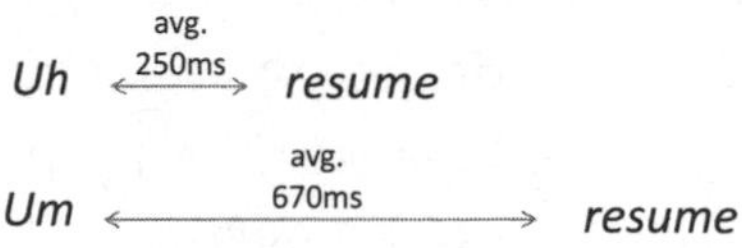

avg.
Uh ← 250ms → *resume*

avg.
Um ← 670ms → *resume*

Figure 5.1 Different average time delay before resumption of speech after "uh" versus "um." Data from Clark and Fox Tree 2002.

speech resumed. They found that if a speaker marks a delay with "um," that delay will be longer than if it had been marked by "uh." If "uh" is used, the delay that follows before resumption of fluent speech is roughly a quarter of a second; if "um" is used, the anticipated delay approaches three-quarters of a second (see Figure 5.1).

They concluded that the meanings of these two words can be defined as follows:

> "um" = I am initiating what is expected to be a major delay in speaking

> "uh" = I am initiating what is expected to be a minor delay in speaking

One might think that these results reveal a fascinating little quirk of the English language, but they are more than that. With "um" and "uh," English supplies its speakers with dedicated ways of indicating what kind of delay is to follow. The delays that are signaled by "um" and "uh" are often caused by problems of processing—recalling words or names, gathering one's thoughts, trying to anticipate others' interpretations—but when people use these words, they are not always experiencing real processing problems. Like any ritualized form of communication, these little words can be used for secondary purposes. As the sociologist Emanuel Schegloff has argued, some uses of "um" and "uh" may not have anything to do with trouble at all.[6]

One use for "um" and "uh" is at the beginning of turns that are dispreferred responses. We saw in the last chapter that dispreferred responses are typically delayed, and we also saw that this delay is not only created by inserting a silence, but also by inserting audible alternatives to beginning the turn, including "um" and "uh." Here is an example (with "uh" underlined):[7]

Stan: Are you registered at your new address?

Joyce: No

Stan: You wanna be registered there? Er at nine two five oh—

Joyce: No because I'm probably moving in June

Stan: Okay yeah that's good

Joyce: You know and then I'll just have to—

Stan: Any changes of uh party affiliation or anything like that?

Joyce: <u>Uh</u> not at this moment. When do I have to tell you by?

In the last line, Joyce is answering Stan's question with a response that appears to resist his line of questioning. This is the kind of case we saw above, in which the presence of "uh" serves to delay the dispreferred response. Here is an example:[8]

Mike: Is that his wife who works there sometimes too?
Sha: <u>Uh</u> no it's not. It's another girl.

This is another straightforward case of a dispreferred response—"no" to a yes/no question—being postponed by inserting the explicit signal "uh."

Schegloff gives other kinds of examples that are dispreferred and that, accordingly, are prefaced with "um" or "uh." The dispreferred types of turns we have looked at so far in this book have mostly been responses to questions. They respond to something that somebody else launched in the conversation, and they are dispreferred because they somehow fail to comply with what the person was going for. Another kind of dispreferred move in conversation is of the kind that launches a new line in the conversation, especially when this implies some form of imposition on the other person. See the fourth line in the following example:[9]

1. **Stan:** Well okay that's about all I wanted to bug you with today.

2. **Joyce:** Okay Stan.
3. **Stan:** So are you okay?
4. **Joyce:** Yeah (pause) um (pause) whatta ya doing like late Saturday afternoon?
   (Stan describes how he is planning to catch up with a friend)
5. **Stan:** Why what's happening?
6. **Joyce:** Because I'm going down to San Diego. And I'm going to fly. And so I need somebody to drive me to the airport.

In line 4 of this example, Joyce launches a line in the interaction that is designed to lead to a request to get a ride from Stan. It is an imposition of sorts, which helps to explain why Joyce's first move is to check Stan's availability to help with giving her a lift. Joyce may in fact be having some internal trouble processing the request, as she might be thinking about how best to phrase her request. But then she may simply be packaging her turn in a way that uses the available marker "um" as a form of delay. It is not randomly selected. It is a good choice for marking the delay of a dispreferred move because it *claims* that the speaker is having trouble. But in any case, this extended function of "um" fits with Clark and Fox Tree's definition, as they state it in terms of delay, not in terms of trouble as such (though trouble is a common reason for delay). Both theories imply that "um" and "uh" are associated with delay, the difference being whether the delay is an unavoidable effect of having processing problems or whether it is inserted as a deliberate signal.

A final context in which "um" and "uh" occur is in prefacing the reason for a call. The following is a complete phone call, which is over in 26 seconds:[10]

Susan: Hello?
Marcia: Hi is Sue there?
Susan: Yeah this is she.
Marcia: Hi this is Marcia.
Susan: Hi Marcia how are you?
Marcia: Fine how are you?
Susan: Fine.
Marcia: Um we got the tickets
Susan: Oh good
Marcia: and put them in envelopes you know with everybody's name on them, and a big manila envelope is hanging on the uh Phraterian bulletin board.
Susan: Oh that's great.
Marcia: So they'll be there by y'know before noon tomorrow we'll get 'em up there.
Susan: Okay that sounds good.
Marcia: Okay?
Susan: Okay thanks so much.
Marcia: Okay bye bye.
Susan: Bye.
(End of call)

When Marcia says "Um we got the tickets," there is no reason to think that she is having any trouble saying what she wants to say, nor is this obviously a dispreferred sort of move. She is merely moving into the reason for the call, passing on a piece of information.

Here is another case. After some preliminary joking around (this is from a phone call, so they can't actually see each other), Guy gets to the reason for his call. It is signaled by "uh": [11]

John: Hello?
Guy: Johnny?
John: Yeah.
Guy: Guy Detweiler.
John: Hi Guy how are you doing?
Guy: Fine.
John: You're looking good.
Guy: Great so are you. Great, gotta nice smile on your face and everything.
John: Yeah.
Guy: Hey uh my son-in-law's down and uh thought we might play a little golf either this afternoon or tomorrow, would you like to get out?
John: Well this afternoon'd be all right but I don't think I'd better tomorrow.

Even in emergency calls, despite the urgency, it happens a lot that the reason for the call is prefaced by "um" or "uh":[12]

Dispatch: Radio, Hubbell.
Caller: Uh send an ambulance to uh fifteen oh four Ferry Street. A kid hit by a car.

Dispatch: Radio?
Caller: Uh could you send an emergency squad out to fourteen sixty one east Mound Street please.

Dispatch: Police desk.
Caller: Uh could you uh go to uh eleven twenty five Broadway.

Schegloff states that these occurrences of "uh" and "um" do not mark trouble as such, but simply mark that the speaker is now going to provide a reason for the call.[13] This is especially explicit in this one of Schegloff's examples:[14]

(after an amount of small talk)
Alan: Okay. well the reason I'm calling—there is a reason behind my madness.
Mary: Uh-huh.
Alan: Uh next Saturday night's a surprise party here for Kevin, and if you can make it.
Mary: Oh really?!
Alan: Yeah.

The presence of "uh" here does not necessarily mean that the speaker is having any actual trouble responding.

Schegloff notes that here the use of "uh" is chosen as part of the delivery of the dispreferred response,[15] helping to delay it. These callers are each taking a signal that is associated with processing problems and using it as a tool for achieving a specific communicative purpose.

There are gender and age differences in people's use of "um" and "uh." The linguist Mark Liberman investigated this question by looking at a large corpus of English spoken language.[16] He sampled more than 23 million words, from nearly 12,000 different people in conversation, and counted up how often people said "um" and "uh."

A first observation is that while all people make frequent use of these words, men use them more than women do. In men's speech, exactly one out of every 50 words was "um" or "uh." For the women, it was around one out of every 70 words. He looked closer to see if there were differences between the use of "um" as opposed to "uh." He found that women use "um" more often than men do: For women, in every 100 words, one word is "um," while for men it occurs slightly less often than that. On the other hand, men use "uh" much more often than women do: For men, about one in every 80 words is "uh," while women will say more than 200 words on average before they say "uh" once.

The observations made by Liberman about the different frequencies of these words depending on the gender of the speaker lead to some interesting possible conclusions. One possibility is that men experience cognitive problems that cause a minor delay in speaking more often than women do. But remember that these little words are not just symptoms of cognitive turbulence. They can be purposefully used as a way of *claiming* that one has a reason to delay speaking. They say "Despite my delay, don't take up the floor yet, because I'm not done." So another possible difference between men and women in this regard might be that men are simply more interested in using this tool to hold onto the floor once they have it. If men are exploiting the system to keep hold of the floor, it is interesting that they are choosing "uh" rather than "um" for this purpose. Recall the finding by Clark and Fox Tree that "um" tends to prefigure longer delays, and "uh" shorter delays. We don't yet know why men are using the "short delay" marker "uh" so much more than women do: This is a matter for further research.

While some are of the view that "uh" and "um" are fully fledged words with defined meanings,[17] others regard them as nothing more than symptoms of internal trouble with processing language.[18] A strong argument in favor of the view that "um" and "uh" are words in the English language is the fact that they don't occur in all languages. If they were literally symptoms of human cognitive processing problems causing delays in speech production, then they should surface in the same or similar form globally, just as abnormal dilation of one pupil can be a symptom of brain injury, no matter what language you speak. There is no language whose speakers do not experience problems and delays in speaking, but there are many languages that use sounds other than "um" or "uh" to signal these delays. Here is just one example, from Lao, the national language of Laos. In Lao, people don't say "um" or "uh" when signaling a delay in speaking. Instead, they say *un* (rhyming with "un" in the English word "under"). This proves that the words do not just come naturally out of people's mouths, but instead people have to learn to make the sound that is conventional in the language being spoken.

Let us take stock of what we have learned from the study of "um" and "uh." These little words are among the most widely relied upon signals in language. They mean something like "I need a moment while preparing what I want to say, so expect a brief delay; I'm not ready to yield my turn at talking." As simple as this message seems, it takes us to the core of some defining elements of the conversation machine and points to what is especially human about the way conversation is organized.

First, these little signals are reflexive. They show the curious property, unique to human language, of being part of a communication system that can communicate about itself. They don't add information about the current topic of conversation; they add information about the state of mind of the speaker and about the flow of the conversation itself.

Second, these signals make sense only in the context of a high-speed turn-taking system. In this system, the smallest delay leaves a speaker vulnerable to losing the floor to someone else who might start talking. And a delay is accountable in this morally committed form of joint action. Given the extreme sensitivity to time that exists in conversation, delays invite all manner of inference—for example, that the speaker is hesitant. With signals like "um" and "uh," a speaker can explicitly signal that they have things under control.

A third thing is that when people use "um" and "uh," they assume that their listeners are cooperative in that they are willing to abide by the wait signal and refrain from jumping in when it might otherwise appear that they could.

The signals "um" and "uh" occur very frequently—too frequently, many would say—in spontaneous speech. They occur regardless of whether the context is a monologue such as a narrative or the to and fro of conversation. Let's now consider a similar type of expression, but one that is exclusively found in the to and fro of conversation. This is the expression "mm-hmm" (or its open-mouth equivalent "uh-huh").

One use of "mm-hmm"/"uh-huh" is with the meaning "yes" in response to a question. Here we are not going to look at that function but rather at its function as what has been termed a continuer in conversation. Look at the following example (with "mm-hmm" highlighted):[19]

Caller: This is in reference to a call that was made about a month ago.

Host: Yes sir?

Caller: A woman called, uh sayin she uh signed a contract for her son who is—who was a minor.

Host: Mm-hmm.

Caller: And she claims in the contract there were things given, and then taken away, in small writing.

Host: Mm-hmm.

Caller: Uh now meanwhile, about a month—uh no about two weeks before she made the call I read in—I read or either heard—uh I either read or heard on the television where the judge had a case like this.

Host: Mm-hmm.

Caller: And he got disgusted and he says I—he's sick of these cases where they give things in big writing and take 'em—and take 'em away in small writing.

Host: Mm-hmm.

Caller: And he claimed the contract void.

Host: Mm-hmm.

Caller: Uh what I mean is it could help this woman that called.
You know uh, that's the reason I called.

The scenario should be familiar. One speaker is talking at length, and as the other listens, one of the ways in which they display that they are paying attention is to interject with "mm-hmm"/"uh-huh" at various intervals. You may be thinking that people are capable of saying "mm-hmm"/"uh-huh" at various intervals when they are in fact paying no attention at all to what is being said. This only lends support to what I am saying here about the function of "mm-hmm"/"uh-huh." If it says that the speaker is paying attention, then this explains why people would use it to try to give the *impression* that they are paying attention. This gives further support to the view that these words are parts of real language and not uncontrolled symptoms or by-products. The same is true for "um" and "uh," discussed above. They share with regular words the property that they can be used to lie and mislead.

When a person says "mm-hmm"/"uh-huh," they are effectively saying to the other something like "please continue" or "do go on, I'm listening." This is clear from the example above. In a sense, this is what "mm-hmm"/"uh-huh" means. But there are more specific accounts of its meaning.

Psychologist Herbert Clark has proposed that "mm-hmm"/"uh-huh" works as a continuer because it answers "yes" to an implicit question that comes with every utterance, namely "Do you understand me and accept what I am proposing up to this point?"[20] Sociologist Emanuel Schegloff argues a slightly different point. He says that "mm-hmm"/"uh-huh" means something like "There is nothing in what you have said to this point that requires me to ask for repetition, clarification, or correction."[21] While subtly different, these two versions of what "mm-hmm"/"uh-huh" means both suggest that it provides useful information in conversation. In the example above, it seems like the speaker who is relating their experience might find it handy to be updated that the listener is following them.

It turns out that the effect of producing continuers like "mm-hmm"/"uh-huh" is significantly more than just handy to know. Clark highlights the importance of the little signals that people make when they are listening to others tell narratives: "Narratives seem different from conversations, because they seem to be produced by individuals speaking on their own. . . . But appearances belie reality. Narratives rely just as heavily on coordination among the participants as conversations do. It is simply that the coordination is hidden from view."[22]

Psychologist Janet Bavelas and her colleagues devised an experiment to study the effects of "mm-hmm"/"uh-huh" in its role as a signal of feedback in conversational narratives. They looked at how people behave when they are listening to someone tell them a story in conversation. They asked people to visit their lab and narrate near-death or near-miss experiences.

These kinds of narrations have a few defining properties, some of which we encountered in Chapter 2. First, they will take multiple turns, so one party will have to listen and remain mostly quiet for the duration of the narrative. Second, the number of turns in which the story will be completed is not determined in advance. Third, the narrative will have a clear point of completion, for example, in the experiments by Bavelas and colleagues, the point at which it becomes clear that the narrator had a brush with death but the danger was averted. And fourth, for the narrative to achieve its true resolution, its completion must not only be recognized by the listener, but the listener must produce a signal of appreciation of the narrative, which should acknowledge the story's meaning and, ideally, should validate the speaker's sentiment by expressing the same evaluation of what has been described in the narrative. For example, if the narrative was a near-death experience, an appropriate way to react at the end of the narrative might be something like "Oh my gosh you were so lucky," with appropriate tone of voice and facial expressions of horror or pity.

These properties of narratives implicate both the speaker and the listener. They invoke rules in conversation of the kind that we discussed in Chapter 2. These rules require *both* participants in a conversation to collaborate in specific ways if the narrative is going to work. Misfires are easy to imagine. A narrator who never gets around to the punch line will be open to rebuke, as might a listener who simply stays silent at the completion of a story or indeed who blurts out "Oh my gosh you were so lucky" at a point when the narrator is just setting up their preamble to the key events. These points imply that there is an especially close relationship between the behavior of speaker and listener, where their contributions interlock, just as the concept of a conversation machine would imply. A narrator needs a listener if they are going to produce a narration that has the required elements.

Bavelas and her colleagues used the methods of experimental psychology to examine just how closely integrated people are in conversation.[23] They brought people into their lab and recorded them in conversation, in pairs. They assigned the role of narrator to one person and asked them to tell the other person about a near-death experience, or an experience in which they had a lucky escape. The experimenters noticed that listeners to these narratives would give two kinds of response during the course of the narrative, which they labeled generic responses and specific responses.[24] Generic responses include things like nodding and saying "mm-hmm"/"uh-huh," as in the following two examples, taken from their recorded narratives:[25]

Narrator: We stayed in an RV park.
Listener: Mm-hmm (with nod).

Narrator: I have a single bed
Listener: (nod)
Narrator: with a headboard.
Listener: Mm (with nod).

Generic responses do not express any meaning that is specific to the content of what the narrator is saying. They merely imply that the listener is paying attention and the narrator may continue.

Specific responses express a sentiment that is connected to the content of what the narrator is saying. Here are three examples (the first two are nonverbal):[26]

Narrator: He flipped his truck over, over an embankment.
Listener: (facial display of concern)

Narrator: No one was around, and he said "Get in the car."
Listener: (facial display of fear)

Narrator: I, like an idiot, decide to climb up the cliff instead of . . .
Listener: . . . going up the road
Narrator: . . . taking the easy way out and going up the road.

With specific responses like these, a listener demonstrates the dramatic or emotional significance of what the narrator is saying.

Both kinds of response behavior place real demands on listeners, who are committed to closely monitoring the progress of the narrative and choosing relevant moments to produce one or the other kind of response. Bavelas and colleagues noted that when listeners produce these responses, they "become, for the moment, co-narrators who illustrate or add to the story." Their hunch was that narrators cannot tell stories well unless they have a listener not only to cooperate but to collaborate in telling the narrative. They suggested that the behavior of listeners was so integral to the success of a speaker's narrative that "distracting the listener from the narrative should affect the quality of the storytelling."

To test this, Bavelas and colleagues gave different instructions to listeners to these near-miss stories. Some were instructed just to listen, while others were given other tasks (unbeknownst to the narrators). For example, some of the listeners were required to count in their mind—while the narrative was being told to them—the number of days, and holidays, remaining before Christmas. This distracted their attention from the narrative completely. Others were required to pay close attention to the narrative, but for a very different reason: They were asked to press a button, located underneath the desk, whenever the narrator used a word starting with the letter "t."

This had two striking effects. The first was an effect on the behavior of listeners. When listeners are distracted from the content of what is being said in a narrative, they produce far fewer responses, especially of the specific type, which express a sentiment that aligns with the content of what the narrator is saying. In the normal narrative condition, listeners produced one specific type of response every 27 seconds, while in the t-counting condition these responses almost disappear, going down to a rate of one specific type of response every 12.5 minutes.[27]

The second was an effect on the behavior of narrators. Bavelas and colleagues found that the fluency and quality of a narrator's speech are worse if their audience is distracted. If a listener produces an appropriate assessment, at the appropriate time, then the story will climax and be resolved neatly. But if the listener is distracted, as in these experiments, the narrator does not get the required signals that their work is done, and a few things will happen. One is that the narration will be extended, looping back around to the punch line. Another is that the narration will become more choppy or disfluent, with the narrator producing more signals of hesitation ("uh"/"um"), changing pace, or producing gaps or pauses. And a third effect is that narrators will try to justify their story as really having been a near-miss, as if the reason for the listener's failure to react was that they did not regard the narrative as appropriate or adequately impressive.

In this example the narrator is describing an incident in which he was working as a logger. A tree started coming down into the space where he was working, and he was unable to jump to the side, so he had to run away from the falling tree in an attempt to outrun it. The narrator's story culminates as follows:

> So this tree's falling, falling, falling. And he was ahead of me, and I was behind, and just the end of the tree clipped my foot. And it felt like, like a *whip* hitting my foot.

Here the narrator has held up their part of the bargain in this conversation. They committed to relating a near-miss experience and in return were given the floor for an extended period. They built up the story and brought it to a conclusion. But there is a problem. While the listener sitting across the table is seemingly paying attention to the story, in fact they have their finger on a hidden button and they are concerned only with whether the words they are hearing start with "t" or not. The result is that they miss their cue. They miss the moment at which, according to their part of the bargain, they should have produced a meaningful appreciation of the story—and so the narrator does not get closure. Suddenly he is adrift:

> And so uh after I, I mean, I saw it fall we both go diving into the thing because we knew—I mean, I don't know how exciting that is but afterwards, uh, I mean, we chuckled about it at lunch. Cause it's always funny if you don't get landed on, sure it was a hoot but [stylized laugh]. Um. I just thought that was, uh, that was funny that, uh. Like *usually,* the easy way to go out is to go to either side, and that way it'll fall

> and you're on either side. But since we had no escape route, we knew it was coming at us, so we had to run for our lives basically, which puts a little excitement into the job too, cause it's fun, rappelling down trees and stuff and, and what-not. So . . . that's all!

This rambling crash landing of a narrative is not the fault of a narrator who has gone off the rails. It is a problem in teamwork. These experiments demonstrate that listener responses—as a form of traffic signaling—affect the linguistic performance of narrators. They reveal the conversation machine at work. While we tend to think of a narrative as a monologue, which by definition involves one person, in everyday conversation the listener makes important contributions throughout. These contributions feed back into the performance of the narrator, implying that the two (or more) people involved are hooked up to each other, together creating the conditions for the conversation machine to run.

This reveals a key element of the conversation machine: our uniquely human cognition for social contingency. A first building block for our ultrasocial cognition for conversation is a seemingly simple awareness of, and interest in, contingency between events. A conversation can be thought of as a series of moves, like moves in a chess game. Each move sets the scene for a response and changes the horizon for where we might go from there. Had a different move been made, a different set of responses would have been possible. Each move is contingent on the previous one. Keeping track of these contingencies is crucial to participating appropriately in a conversation.

This sensitivity to contingency and codependency in social interaction is part of a species-wide conversation machine that facilitates human interaction as we know it. Evidence comes from research on social interaction in child development. The psychologists Carolyn Rovee-Collier and David Rovee carried out a simple study with babies, establishing that they are already tuned in to contingency at the age of two months.[28] In their experiment, two-month-old babies lay in their cribs looking up at a colorful mobile hanging over them. Not surprisingly, the babies preferred it when the mobiles were moving around than when they were still. The experimenters created two contexts in which to measure the babies' interest in the moving mobiles. They tied a soft, silk cord around each baby's ankle. For one group of babies, the cord was looped up and tied to the mobile, meaning that the babies could directly cause the mobile to move around, thus reinforcing the fun stimulus. In the other situation, the cord around the infant's leg was not attached to the mobile. The mobile continued to move, but those movements had no direct connection to the infants' leg movements.

In both situations the babies saw a colorful mobile moving around in front of them. Any infant should of course find this interesting. But when that movement is caused by the infants' own bodily actions, not only did infants become more intensely interested, and for a longer time, but the strength and frequency of their leg movements tripled, thus further reinforcing the involvement.

This finding shows that even at two months of age, babies are attuned to their capacity to affect the world around them and to cause things to happen. This ability to affect things is not only fun; it is also the basis of the child's understanding of causation and of their own developing sense of control and agency. Infants quickly tune in to the relation of contingency. They understand the dependency between events—such as a kick of the leg and the gratifying result, the shake of a mobile hanging over their crib.

The relation of contingency between action and reaction is not only important to people for its relevance to acting in the physical world. It is at the core of our understanding of the social world as well. When babies are just two months of age, their social interactions with adults also show a clear contingency in the relation between each side's contribution. Adults and infants do not merely want to look at each other moving and expressing themselves, as one might enjoy looking at a colorful mobile moving in the air. They want to engage in conversation, in which each action by one person is a reaction tailored to the other, which in turn invites a further reaction.

This is real interaction, not just being together, not just being the target of someone's talk and movements, but locking into that other person's behavior in such a way that the two people are part of one thing. This is the insight we noted earlier from philosophers of social action such as John Searle and Margaret Gilbert. For them, joint action is when the people involved are thinking of what they are doing in terms of what "we" are doing, not what "I" am doing. This capacity is at the heart of the conversation machine.

In a classic demonstration of how interaction is a game of joint contingency, developmental psychologists Lynne Murray and Colwyn Trevarthen carried out a study of interaction between two-month-old infants and their mothers.[29] The mother and child were placed in separate rooms, each in front of a full-size, full-face video image of the other. Each could see and hear the other, and they were simply left to interact via the video link. The experimenters created two different situations for the mother-child pairs. For some of them, the video link was live. The mother and infant each saw and heard the actions of the other in real time. For others, instead of seeing a live feed of their child's behavior, what the mothers saw (unbeknownst to them) was a recording of the child that had been made earlier.

In both conditions, the mother and her two-month-old child saw and heard each other. But they behaved very differently in the two situations. In the live-feed situation, the behavior of the babies was directly responsive to the mothers. The immediate contingency of the babies' responses in turn heightened the mothers' degree of attunement to their children. Mothers engaged more in baby talk. They used shorter sentences, more repetition, and more emotive and expressive speech.

In the nonlinked situation, even though the mothers believed that they were interacting live, the lack of clearly contingent response from their babies had the effect of removing the usual attunement between mother and child. In the non-live situation, the mothers talked like they would talk to adults, even though their child was only two months of age. Murray and Trevarthen concluded that the infant was playing an active role in determining how the mothers spoke. The features of baby talk that mothers often produce are adaptations to such a capacity,[30] which in turn explained why mothers' baby talk was subdued in the non-live experiment.

At the time when Murray and Trevarthen carried out their experiment, the prevailing view among researchers was that mother-child interaction was not true back-and-forth conversation. Rather, researchers believed that these exchanges only resembled conversation, because mothers were skilled in "filling in the pauses in the flow of the infant's spontaneous and unresponsive behavior." But if this were true, there would have been little difference between mothers' behavior in the two situations in the video-interaction experiment: in response to the baby on a live feed versus in response to the baby in a recording. Instead, the experiment revealed that the babies' capacity to respond appropriately in the flow of interaction gave structure to the conversation and made it more coherent. It directly affected the way in which the mothers spoke and acted in these interactions.

These experiments by Murray and Trevarthen are a good illustration that both parties to interaction act in ways that are mutually contingent. This is how the conversation machine works. Any move in conversation is at the same time a reaction to what just happened, and something that in turn elicits a reaction next. The mother and child in these simple interactions are doing what adults do in full-blown conversation. They collaborate to produce a joint outcome. Each individual plays their part in a joint activity. By coupling and behaving in ways that show direct contingency and mutual responsiveness, together they form a system as parts of a single machine.

Notice the similarity between this experiment with two-month-olds and the earlier experiment with narratives about people's near-miss experiences. Distracted listeners couldn't produce the normal subtle cues that listeners should produce. This had a direct effect on the behavior of the people who were narrating their stories. They became less fluent. They doubled back and repeated themselves. They tripped over the story climax and found themselves justifying the suitability of their story, when they had in fact already made it clear. The reason these things happen is that stories in conversation are collaboratively told, even if one person's role is as "mere listener." The mothers' reactions to their nonresponsive infants' behavior were similar in kind to those of the narrators who lacked cooperative listeners. In both cases the experimental manipulation tampers with the workings of the conversation machine.

These experiments on both mother-infant interaction and narrator-listener interaction establish that conversation requires more than mere synchronization in behavior. Synchronized action is widespread in the animal world. In dolphin societies,[31] groups of individuals can form close social coalitions. For example, males will form partnerships in order to defend their access to certain female dolphins. They mark these partnerships by imitating each other closely, both in the way they move when swimming and feeding, and in the sounds of their calls. Behavioral ecologist Peter Tyack has drawn an analogy between social signaling in dolphins and a phenomenon in human language known as accommodation. A number of communication scholars have shown that when people converse, they tend to adjust their way of speaking to be more similar to each other.[32] This can involve a wide array of features of behavior. People will tend to imitate each other in terms of how long their sentences are, how fast their speech is, the sound and intensity of their intonational patterns, what accent is used, what words are chosen, how much self-disclosure occurs, and the patterns and frequency of head nodding and body posture.[33]

Of course, being human does not mean we stop being animals, so it is not surprising that we would show similar strategies to other animals in trying to achieve and display interpersonal affiliation. But the accommodation-style forms of interactional attunement just listed—and apparently shared with dolphins, among other species—are different in kind from those that make human conversation unique. People in conversation do not just travel together in step. They create something together. To achieve this, they have to be interdependent. A questioner needs a respondent. A gossiper needs a confidant. A narrator needs a listener.

The interlocking contingency that glues each move together in conversation is clearly a defining part of conversation itself. But contingency between moves in a communicative sequence is not enough. We have also discussed the accountability that comes with participating in conversation as a form of cooperative joint action. The interdependence that makes conversation possible is a defining element of our species' elite capacity for cooperation and for the moral thinking associated with it. Developmental psychologist Michael Tomasello argues that interdependence is the key to human uniqueness in social cognition and social interaction.[34] Interdependence underlies our moral lives as well. If we are going to be interdependent, then we must each commit to playing our part. Not sticking to that commitment would be a moral failing.

It might seem far-fetched to claim that words of the kind we have considered in this chapter—from "um" to "uh-huh" and beyond, to "oh," "so," and "okay"—have a moral architecture. But that is precisely what I want to say. These words are not your regular nouns or verbs.[35] They don't refer to things or describe events. Instead, their function is procedural.

The use of little traffic signals such as "uh"/"um," "uh-huh," and "okay" all illustrate ways in which bits of language are used for regulating language use itself. In addition, they illustrate another core feature of the machine: the norm-governed, cooperative nature of human interaction. These words work because they presuppose that people will agree to follow the instructions that they issue. By using these signals, we are following norms of coherent interaction, and we are also revealing that we are sensitive to the commitments that underlie conversation.

6
THE GLUE OF RELEVANCE

We have seen that transitions between turns in conversation are timed to near-perfection. And as we saw in the last chapter, the traffic signals of conversation can keep the line of talk on track, whether it be by reassuring the other person that a response is still coming despite a delay, or by signaling to them that you have understood their narrative so far. The conversation machine must have ways to carry out these organizational functions in conversation, for the simple reason that conversation is unscripted. When we talk, we collaborate in constructing long, complex, and tightly timed dialogues on the fly. A conversation is not just a duet; it is an entirely improvised duet.

To understand what this means, consider cases of vocal alternation in animals—often also called duetting—that are often compared to turn-taking in human conversation. Numerous species show behavior that resembles conversational turn-taking insofar as one individual makes a sound, and then another follows, before the last one goes again, and so on. An example of this is the marmosets' call-and-wait alternations that we discussed in Chapter 2. Clearly there is no improvisation here, as the call that each marmoset produces is the same one every time. A marmoset doesn't have to compose a new and appropriate response at every turn.

In other primate species, forms of vocal interaction are more sophisticated. The siamang is a species of tree-dwelling gibbon found mostly in Sumatra and peninsular Malaysia. Male-female pairs of siamangs engage in loud and elaborate calling bouts. They are not unlike duets in human song—think Sonny and Cher's "I Got You Babe"—in that the individuals each contribute to a preordained set of alternating moves, one leading to the next, until the end of the planned song has been reached.

Figure 6.1 is an illustration of the structure of the siamang great call sequence. In the diagram, time passes from top to bottom. The two columns represent the male (on the left) and the female (on the right).

There is clearly a pattern of alternation between the two individuals (though with clear overlap), where specific calls made by one individual will be responded to with appropriate calls by the other. One stage of the song leads into the next one, eventually leading to a resolution

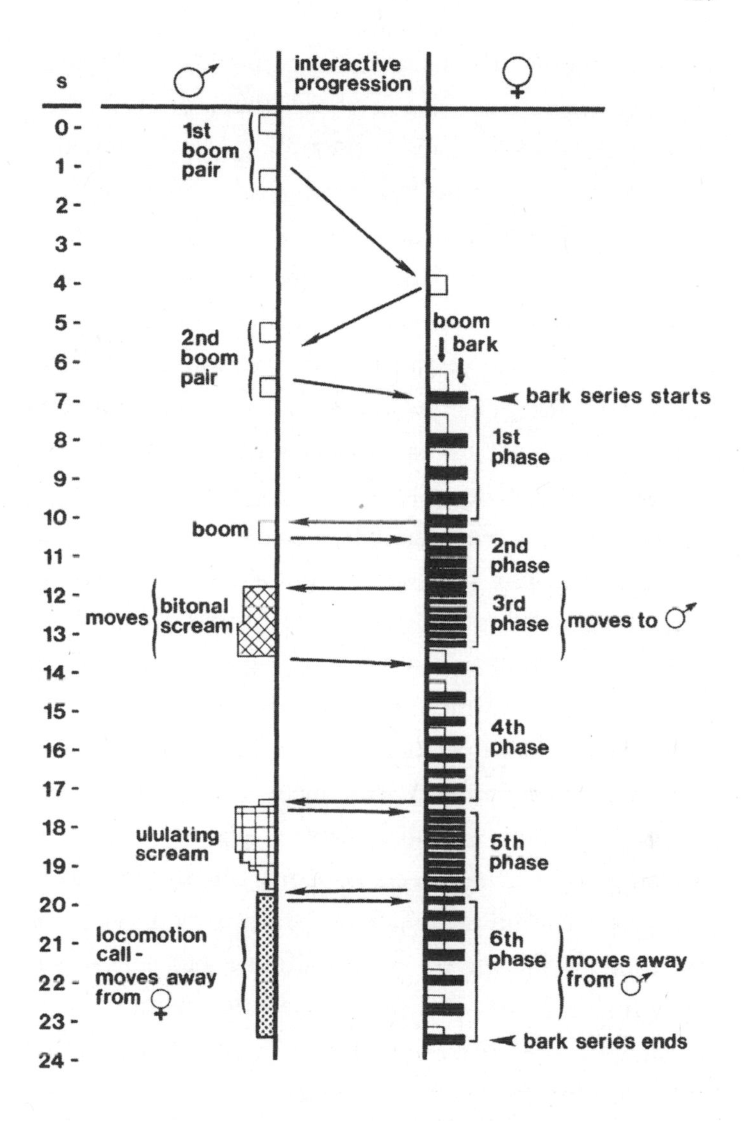

Figure 6.1 A simplified and stylized transcript of the most basic vocal and nonvocal behaviors produced during the latter stages of the organizing sequence and the entirety of the 120 siamang "great call sequences" analyzed by Elliot H. Haimoff in 1981. Source: Haimoff 1981: 141.

and completion.[1] The siamang duets show that two individuals can create a sequence of calls that are closely contingent, not only in terms of the timing of a call but also in terms of the specific call that is made. Each member of a siamang pair must do its part, at the right moment relative to the other, if the two are going to get to the end of the song together. But still this is unlike human conversation. One of the key observations that Sacks and colleagues made about turn-taking in human conversation was that human conversation is not scripted. The fact that we can carry out conversations seemingly without effort is remarkable because we do not know in advance how long the conversation is going to last and what the people are going to say, or in what order.

If a required move in a siamang duet is missing, the interaction is simply aborted. This shows that these lesser apes are far from doing turn-taking as it is known in human conversation. When misfires happen, such as a failure to produce an appropriate vocalization at the expected moment, the wheels just come off. But in human conversation, things are kept on track in two key ways. First, people show a constant willingness to hold others accountable if they contravene the rules of interaction. Second, as we shall see in the next chapter, if something is unclear or needs repeating, then a dedicated system of repair is used. We can point out to each other when something is missing, we can sanction inappropriate moves, and we can pursue the responses that we were expecting or hoping for.

The closest primate species to ourselves—the great ape species, chimpanzees, bonobos, gorillas, and orangutans—do not display vocal turn-taking at all, not even of the kinds seen in marmosets and siamangs. However, they do have sophisticated forms of interaction that show clear contingency between moves in a behavioral sequence, and in which individuals may pursue the responses they are expecting. These forms of interaction seem much closer to our form of interaction in that they are sequentially organized and each move is a meaningful response to the prior one.

Cognitive scientist Federico Rossano studied interactions among bonobo mothers and their infants, focusing on the ways in which infants get their mothers to pick them up.[2] Figure 6.2 is an illustration of one of these interactions, between an infant, Fimi, and her mother, Yasa.

The figure shows three simple phases. First, Fimi looks to her mother (Figure 6.2a) and waits until the mother is looking at her. Second, once the mother is looking, Fimi then makes the "wrist bent" gesture (Figure 6.2b), a conventional gesture that bonobo infants use

Figure 6.2 Three-phase sequence in which the infant bonobo Fimi uses the "wrist bent" gesture, directed to her mother, Yasa, as a request to be picked up. Source: Rossano 2013: 167.

for requesting that their mothers pick them up. Finally, the mother responds by picking up Fimi and carrying her (Figure 6.2c).

Rossano argues that these sequences are organized in ways that are similar to human conversation. A key to his argument is that if bonobo mothers do not respond in the desired way—i.e., by picking the infant up—then infants will continue to pursue the response they were seeking, for example by repositioning and trying again. Compare this to the siamang case, where a missing or unexpected move simply puts an end to the entire sequence.

Also note the timing properties of the bonobo interactions. The time that passed between Fimi making the bent wrist gesture and Yasa moving to pick her up was brief. Rossano timed it at less than 200 milliseconds, the length of the average transition of turns in human conversation. Rossano documents numerous sequences of this and similar gestures, and observes the same timing properties.

Human conversation has something genuinely in common with the bonobo request exchanges, but only superficially in common with the vocal exchanges of marmosets and siamangs. The point of similarity with the bonobo interaction has to do with the *meaning* that links each move to the next. The key point here is that human conversation is an exchange of co-relevant moves, not an exchange of simple or planned calls. In the back-and-forth of conversation, people don't do the same thing in every next turn (unlike the marmosets), nor do they do a set thing in next position (unlike the siamangs). Something in the conversation machine not only drives the fine timing of conversation; it also provides for meaningful connections between each move that allow us to create coherent conversations without a script. This connection comes from a special glue called relevance.

Relevance is one of the most powerful cognitive components of the conversation machine.[3] There is a deep-seated human propensity to see meaningful connections between events, whether or not the events are actually related. We find it hard *not* to see events as being meaningfully connected when in fact they are simply adjacent. Many superstitions work on this basis: If I stub my toe, I might see this as connected with the black cat that just walked across my path.

In natural conversation, people constantly exploit the principle of relevance. Cognitive scientists Dan Sperber and Deirdre Wilson give the following example:[4]

Peter: Do you want some coffee?
Mary: Coffee would keep me awake.

Mary's turn occurs immediately after a yes/no question from Peter. Human cognition is wired to see relevance in this response, despite it not being a direct answer to the question. We cannot help but assume that the turn is directly relevant to Peter's turn. This is not only because Peter has just said something to her, but also because Peter's turn was a direct question, and this set up a moral duty for Mary to respond. Under these conditions, we assume that Mary's turn is relevant and that it is an answer to Peter's question/offer.

Of course, precisely how we interpret Mary's response depends somewhat on the context. If it is obvious that Mary does not want to stay awake—e.g., if it is late at night—then we interpret Mary's response to Peter's turn as being equivalent to an answer of "no." Equally, it may be obvious that she *does* want to stay awake, in which case the turn means "yes." Notice, by the way, that Mary here has a way of avoiding saying "no" explicitly and thus not have to deal with the dispreferred nature of rejection (discussed in Chapter 4). She does this by simply stating the reason for her declination and relying on Peter to infer what she is really saying. If he assumes her response is relevant, then it will be relevant.

On the siamang duet model of social interaction, the interaction would come to a halt if Mary provided any answer other than the "yes" or "no" that the question asked for. But people in conversation will do their utmost to interpret each move as being relevant to the one prior. The human assumption of relevance, supported by our higher-order social intelligence, will incorrigibly glue each move to the next, even when there is no true connection whatsoever. The sociologist Harold Garfinkel summed up this point by remarking that people will always understand what you say and do, just not always in the way you intended.

A vivid demonstration comes from a 1960s experiment that Garfinkel conducted in the Department of Psychiatry at UCLA.[5] Participants in the experiment were told that they were going to experience a new form of counseling therapy. They were asked to sit in a treatment room and were told that their counselor would be seated in a separate room. They were to communicate with the counselor via microphone only. Their instructions were that they should pose only yes/no questions. The new therapy being tested allowed the counselor only to give "yes" or "no" responses. The participants were free to talk for as long as they wanted when giving background to each question.

Participants were unaware of a major twist in the design of the experiment. The "yes" or "no" answers that the counselors would give bore no relation whatsoever to the questions that the participants asked. What the participants did not know was that the counselors had a written list of "yes" and "no" answers. The counselor would give each answer simply by reading out the next word on the list.

Garfinkel found that participants in the experiment seldom caught on to the fact that the answers bore no relation to their questions, even when the setup led to complete contradictions. In some cases, a participant would ask the same question at two different points in the consultation. By chance, they sometimes got two different answers.

In one instance, the participant was a male Jewish student who had recently begun dating a female Gentile student. His line of questioning with the counselor concentrated on the student's father's implicit disapproval of him dating the girl. He explains that he likes the girl but is ill at ease with his father's disapproval of the situation. Then his question:

> **Student:** Do you feel that I should continue dating this girl?
> **Counselor:** My answer is no.

Later in the session, the student imagines that his father has said to continue dating the girl but still gives the impression that he does not approve. He again poses basically the same question as earlier:

> **Student:** Should I still date the girl?
> **Counselor:** My answer is yes.

While in this case the student expresses surprise at the "yes" answer, he assumes that it is a genuine answer and that it reflects complex reasoning by the counselor.

All the participants interpreted the responses as answers to their questions, that is, as relevant rather than random. Garfinkel's rich discussion of his findings defines people's incorrigible assumption of relevance and the resultant connectedness between moves in conversation. The same assumption of relevance shows up in cultural practices of divination observed in cultures around the world.

In divination, people consult natural phenomena for answers to their human questions. The anthropologist David Zeitlyn has analyzed divination practices among the Mambila, an indigenous group of the Nigerian/Cameroon borderlands in West/Central Africa. The Mambila have a conventionalized practice of spider divination. When a villager has a problem about which they require advice, they may consult a spider diviner. As in Garfinkel's UCLA experiment, Mambila spider divination centers around yes/no questions:[6]

> A hole in the ground inhabited by a spider is covered by an enclosure, usually an inverted pot. A stick and a stone are placed within this enclosure, near the entrance of the spider's hole. A set of marked leaves is placed over the entrance to the burrow. When questions are posed the pot is tapped; in response to the knocking the spider emerges from its hole. In doing so it disturbs the leaves. The resulting pattern of the leaves in relation to the stick and to the stone is interpreted as an answer to a question.

The Mambila villagers, just like the UCLA students in Garfinkel's experiments, take random events and interpret them as rational responses to questions. And as in the UCLA experiments, Zeitlyn reports that when the Mambila are faced with contradictory answers from their spiders, they do not abandon the conversation. Nor do they conclude that it is incoherent. Instead, they see relevance and reason behind it.

The direction of everyday conversations is of course not determined by random lists of words or by leaf-disturbing spiders. But we still apply a strong assumption of relevance at every step. Without such an assumption, it would be hard to explain how it is that humans are so good at seeing the many hidden meanings that underlie the things we say, especially given that human language gives us essentially infinite possibilities for expression.

Indeed, we can think of a million ways in which a person could effectively convey "no" in response to the question "Do you want some coffee?" In the example given above, Mary's way of doing this—"Coffee would keep me awake"—works only because she assumes that Peter will use his head, and in a sophisticated way. This is the nature of language in conversation. While all animals may have reasons for action, only human language allows us to make explicit reference to those reasons, such that others may make inferences about what we really want to say.

In the case of the preliminaries to questions, favors, and so on, discussed above, people are able to see their relevance right away. In the following example,[7] Eddy's query in line 1 is a preliminary to a possible favor, and while he never gets to ask the favor, it is clear to Mike what was coming:

1. **Eddy:** I was wondering whether you were intending to go to Swanson's talk this afternoon.
2. **Mike:** Not today I'm afraid I can't really make this one.
3. **Eddy:** Oh okay.
4. **Mike:** You wanted someone to record it didn't you.
5. **Eddy:** Yeah (laughs).
6. **Mike:** (Laughs) No I'm sorry about that.

The sort of inference that Mike applies here is effortless for people in conversation. On the surface, Eddy merely asked a question about whether Mike would be going to the talk. Yet in conversation we are always asking ourselves *why* people are saying the things they say. If we have a sense of where people are headed, this often allows us to abbreviate conversation in highly efficient ways.

The psychologist Herbert Clark explored this issue in a study of how people can respond in more than one way to a turn in conversation. Clark gives this example:[8]

Is Julia at home?

Clark points out that people can understand the relevance of this question in more than one way. It might be a simple information question, seeking a "yes" or a "no." The questioner just wants to know if Julia is at home. Or on the telephone it might be an indirect way to get someone to bring Julia to the phone. This second reading is possible because the listener is tuned in to the relevance of the question. Julia being at home is a precondition for the possibility of going and getting her, just as Mike going to Swanson's talk would be a precondition for him being able to record the talk. If the listener recognizes this relevance, they can save time and trouble by responding to the turn not as a question but as a request, for example by saying "I'll just get her."

Clark conducted experiments to see how systematic people were in this respect. In one of these experiments, a researcher phoned restaurants in Palo Alto and asked "Do you accept credit cards?" One way to respond to this question is with a simple "yes" or "no." But Clark reasoned that people are likely to use their higher-order reasoning, under the assumption that a caller will have a hierarchy of goals: "As one goal, she wants to decide whether or not to eat at the restaurant, probably that night. As a subgoal, she wants to know how to pay for the meal. As a subgoal to that, she wants to know if she can pay with any of the credit cards that she owns. As the next subgoal, she wants to discover whether any of the cards acceptable to the restaurant matches any of hers."[9] So rather than merely saying "yes" to the question of whether the restaurant accepts credit cards, it is more helpful, and more efficient, not only to confirm that the restaurant takes credit cards but also to state which credit cards are accepted. Indeed, Clark found that in well over half of the cases,[10] the restaurateur would immediately supply the information about *which* cards they accepted, even though the caller had not in fact asked which cards. Most strikingly, Clark found that when the restaurant took only one type of credit card, they would immediately supply this information in 100 percent of cases:[11]

Caller: Do you take credit cards?
Restaurateur: We take American Express.

This little exchange is deceptive in its simplicity. We effortlessly perceive a connection between these turns because we are viewing it through our relevance goggles. The caller asked a yes/no question. Going by a literal interpretation of their words, the restaurateur does not directly answer it, but we perceive that they did in fact answer it especially helpfully and efficiently.

The restaurateur applied two elements of their social cognition that are at the heart of the conversation machine. The first is their ability to apply higher-level mind reading, reasoning about the likely goals of others, goals that can be at some remove from their surface behavior. ("This person is only asking if we take credit cards, but I infer that they will also want to know which ones.") The second is their willingness to cooperate and produce a response that is both efficient and helpful.

This willingness to help is in line with the other-oriented instincts that distinguish human social cognition from that of even our closest relatives in the animal world. Behavioral scientist Alicia Melis and colleagues conducted an experimental comparison of sharing in human children and in our nearest evolutionary neighbors, chimpanzees. The experiments demonstrated ways in which human children from the age of five cooperate and benefit by taking turns while, by contrast, "chimpanzees do not take turns in this way, and so their collaboration tends to disintegrate over time."[12]

Related research shows that while our general cognition—spatial understanding, categorization, causal reasoning—is not unique compared to chimps and other apes, our *social* cognition is what sets us apart.[13] With our capacities for understanding others' minds, for tracking the complexities of social relationships in large social networks, for cooperating with others toward shared goals, and for punishing others who don't play by the rules, humans stand out. This is why real conversation cannot occur in other species.

Sociocultural cognition—including the cognitive capacities for mind reading, relevance, and morally grounded social commitments—is at the core of the conversation machine, and it makes the crucial difference for language in our species. Humans are especially attuned to other minds and to the cultural construction of group-specific, conventional systems of meaning and practice as shared frameworks for communication and joint action. This is what makes it possible for human populations to foster the historical development of complex systems of shared cultural tradition, of which language is one form. If we are going to sustain community-wide systems of linguistic and cultural convention, we need ways to achieve a high degree of intersubjectivity. For that intersubjectivity to exist, people must actively work to maintain it at the micro-level. To do this, people need a way to stave off the constant possibility that common understanding may break down. With each move in our constantly forward-feeding unscripted dialogues, the line of talk can shift to a new direction. If you miss what was just said, or if you are unsure of its relevance, the conversation machine had better provide a system for solving the problem right there, right then, before the moment is gone. That system is called repair, the topic of the next chapter.

7
REPAIR

In the nonstop flow of turn-by-turn conversation, there are many sources of noise, distraction, and unclarity. The pace is fast. Each new turn can take the line of talk in any direction. So, if there is a problem with understanding or hearing what someone just said—a wrong word, a misrecognized name, or an interfering noise—it needs to be resolved immediately, or the chance to resolve it may be gone.

When a fleeting problem of understanding in conversation is identified and corrected, there are two key parts to the process. One part is the initiation of repair, the signal that a problem needs to be repaired. The other part of the process is the repair itself, the way in which the problem is actually resolved. Sometimes a single person is involved in the entire repair process, producing both the initiation of repair and the repair itself. This is self-initiated self-repair. (We saw an example of this in Chapter 5: *First a bro—uh a yellow and a green disk.*) Sometimes it is the *other* person who initiates repair. Consider these two examples from phone calls in English:[1]

A: Dippert's there too.
B: Huh?
A: Dippert is there too.
B: Oh is he?

A: Oh Sibbie's sistuh had a baby boy.
B: Who?
A: Sibbie's sister.
B: Oh really?

The pattern is familiar. Person A says something. Then Person B, instead of moving the conversation forward with something new, draws attention to a problem with what was just said. Person A has to go back and fix it.

In the two examples above, it seems clear what the problem was. Person B just didn't hear clearly what was said. All Person A had to do to solve this was to repeat part or all of what they said, with a slightly clearer, more careful articulation. Once the relevant thing is repeated, we see that Person B is then able to move forward with an appropriate next move in the conversation. In both the examples, Person B gives what is called a news receipt. Often beginning with "Oh," news receipts in conversation make it clear that a person has understood the new information that was just offered, and they tend to invite further elaboration.[2]

These examples with "Huh?" and "Who?" show that the two parts of the repair process—the initiation of repair and the repair itself—can be produced by two different people. This clearly demonstrates that conversation is a cooperative form of joint action. The job of repair is shared. Just as, when making a cake together, one person pours in flour while the other stirs the mixture, here one person draws attention to the problem, and the other repairs it.

The pattern illustrated in the "Huh?" and "Who?" examples is referred to as other-initiated repair. When Person A first says "Dippert's there too" in that example, he may not have been aware of any problem. Person B is the one who initiates repair. By drawing Person A's attention to the problem, he gives him a reason to correct it, and in turn Person A cooperates by obliging.

This is no rare occurrence. With a team of research colleagues, listed in Table 7.1, I carried out a study of repair in conversation in twelve languages from around the world.[3]

We found that other-initiated repair is happening all the time. Our global sample of more than 2,000 of

Language	**Where spoken**	**Researcher**
Cha'palaa	Ecuador	Simeon Floyd
Dutch	The Netherlands	Mark Dingemanse
English	United Kingdom	Kobin Kendrick
Icelandic	Iceland	Rosa Gisladottir
Italian	Italy	Giovanni Rossi
Lao	Laos	Nick Enfield
LSA	Argentina	Elizabeth Manrique
Mandarin	Taiwan	Kobin Kendrick
Murrinhpatha	Northern Australia	Joe Blythe
Russian	Russia	Julija Baranova
Siwu	Ghana	Mark Dingemanse
Yélî Dnye	Island Melanesia	Stephen Levinson

Table 7.1 Languages included in our global comparison of "repair" in human conversation. For each language, researchers collected, transcribed, and coded around 4 hours of spontaneous conversation, resulting in nearly 50 hours of directly comparable material from everyday conversation around the world. Source: Dingemanse et al. 2015.

these sequences showed that one such sequence occurs, on average, once every 84 seconds in informal conversation. This tells us two things about human language. The first—not surprisingly—is that language is far from infallible. Hardly a minute goes by without some kind of hitch: a mishearing, a wrong word, a poor phrasing, a name not recognized. The second is that people do not want to let these problems pass. The high frequency of other-initiated repair shows that people find it important to take the trouble to point out these problems and to solve them as they happen.

The "Huh?" and "Who?" examples given above show two distinct approaches that a person can take to initiating repair in conversation. One strategy is more specific, or stronger, than the other. When Speaker B draws attention to the problem with "Who?," they are being more specific about where the problem is than if they had said "Huh?"

In the "Sibbie's sister" example, the problem turn was a full sentence, an announcement of news about a person, that she had a baby boy. By responding with "Who?" Person B is doing two things. First, they are making it explicit that there was a problem with the reference that Speaker A made to a person. However, note that they are not making it clear exactly what the problem is. It may be that they didn't hear clearly, or it may be that they heard fine but just didn't recognize what was intended, for instance if Sibbie had more than one sister.

Second, they are implying that there was *no* problem with the rest of what was said. When Person B says "Who?," they make it clear that they heard and understood that someone had a baby boy. This strategy allows Person B to be appropriately specific in making the correction. As we see in the example, there is no repetition of "had a baby boy." There is only a repetition, pronounced a little more carefully, of the reference to a person, "Sibbie's sister." The next step in the conversation shows that this has solved the problem. The news is now registered ("Oh really?"), and the conversation can move on. Thanks to Person B's care in being specific, the disruption to the conversation was minimized.

However, saying "Huh?" (or its equivalents "What?," "Pardon?," etc.) is as nonspecific as one can get. It gives the other person no information about where in particular the problem is. Another way to describe this difference in specificity is to call it a difference in strength. "Huh?" is a weaker way of initiating repair. It has less power to direct others' attention to what needs fixing. Accordingly, in the "Dippert" example above, Person A repeats everything they just said, this time with slightly clearer articulation: from "Dippert's there too" to "Dippert is there too." Again, this enables the next step in the conversation to show that the problem has been solved. The news is registered ("Oh is he?"), and the conversation can move on.

We have distinguished between two broad categories of repair initiator: weak and strong. "Huh?" is referred to as weak because it is completely general regarding what the problem was in the original line. It is fitting, then, that in cases like the Dippert example, repair is done by repeating the entire line. An alternative way in which the Dippert example could have played out is as follows:

A: Dippert's there too.
B: Who?
A: Dippert.

"Who?" is a stronger form of repair initiator. It conveys that Person B heard and understood a good deal of what was said, just not all. Here we understand that Person B heard that somebody is there too, but they either did not hear, or did not understand, who it was.

Now let us zoom in on the idea of a strong repair initiator. There are two types. "Who?" is an example of the type of repair initiator that requests specific information, in this case information about a person being referred to. Other question words, such as "When?" and "Where?," are also used in this way. A second type of strong repair initiator does not ask for specific information; rather, it supplies a version of what the person thought they heard and asks for simple confirmation of whether this is correct. As this next illustration shows, if all goes well, using this type of repair initiator allows Person A to solve the problem in the simplest way seen so far: by simply confirming that they had indeed been heard or understood correctly.

A: Dippert's there too.
B: Dippert's there too?
A: Yeah.

There is an important difference between the weak strategy and the two strong strategies. This has to do with the options that people have at any given time in a conversation. Consider the first line in the examples just given:

A: Dippert's there too.

If we think of the flow of conversation as a set of choice points, here Speaker B has two broad options (see Figure 7.1). One of these options is to produce a next move that proceeds with the flow of the conversation and that indicates, or at least claims, that the last line was understood. For example:

A: Dippert's there too.
B: Oh is he?

The other broad option—which is, in principle, always available—is to produce some form of other-initiation of repair. But the different forms of repair are not equal. Of the weak versus strong types, only the weak one is in principle always an option. If, as described above in relation to trouble-prone contexts, one did not hear any of what was said, the best one could do is say "Huh?" or an equivalent such as "What?" or "Pardon?" Only in the case of the strong type is one necessarily selecting that type by choice.

If conversation is a collaborative activity, then it would seem preferable for people to avoid disrupting it.[4] Initiating repair not only momentarily halts the progress of a conversation; it also requires people to go back and re-do a step. This means that repair is a disruption, albeit a low-cost one, to the flow of the interaction.

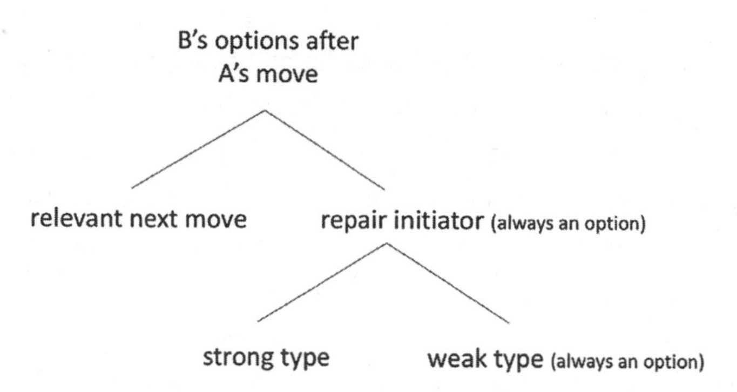

Figure 7.1 The possible ways of responding to any given move in conversation.

Initiating repair in these ways can also be thought of as an imposition on the other person. So, when it is necessary to get someone to repair what they just said, people should be motivated to minimize the trouble that they ask the other person to go to in doing the repair. Now notice the difference between the "Huh?" and "Who?" examples in this respect. In the "Who?" case, the problem can be resolved more economically, with less effort: There is no need to repeat the whole line; just the name will do. But in the "Huh?" case, the whole line must be repeated. This imposes more work on Person A.

These options mean that people can adopt different strategies when doing other-initiation of repair. One strategy is to simplify one's own job and make the other person do the work. This would be the "Huh?" strategy: Don't bother giving the other person specific information; just say "Huh?" anytime there is a problem, and they will have to repeat or rephrase what they said, thus solving the problem. A different strategy is to minimize the other person's effort by being as specific as possible when initiating repair. By initiating repair with more-specific or stronger forms, the other person has an easier time determining what the problem was and in turn has to say less when resolving that problem.

Research on conversation has investigated the choices people actually do make when they need to initiate repair in recorded conversation. A pioneering study of repair in English, by sociologists Emanuel Schegloff, Gail Jefferson, and Harvey Sacks,[5] indicates that ways of doing other-initiation of repair have a natural ordering, based on strength. Their idea was that people prefer to use stronger initiators.

Schegloff and colleagues noticed that people sometimes change from using or launching a weaker initiator, and replace it with a stronger one. Here is an example:[6]

1. **B:** How long y'gonna be here?
2. **A:** Uh—not too long. Uh just till uh Monday.
3. **B:** Till—oh y'mean like a week from tomorrow?
4. **A:** Yah.

In line 3, Person B initiates repair, requiring Person A to confirm whether or not their understanding was correct. At the beginning of the line, Person B starts with "Till—" before cutting off and restarting. We can guess that they were going to say "Till Monday?" or "Till when?" Both of these would require some work from Person A to clarify. We can tell that this conversation takes place on a Sunday. The confusion stems from the fact that while the next day is a Monday, people do not normally refer to the next day by name, but instead just say "tomorrow." Person B seems to grasp this partway through line 3, cutting off and rephrasing with a simple understanding check: "y'mean like a week from tomorrow?" This is a very specific and therefore strong form of repair initiation, and it is easily resolved with a simple confirmation "Yah." The example shows that when people change midstream from one form of repair initiation to another, they often upgrade in strength or specificity.

In other cases, Schegloff and colleagues noticed that a more specific marker can be used when a more general one has failed. Here is an example:[7]

1. **Lori:** But y'know single beds are awfully thin to sleep on.
2. **Sam:** What?
3. **Lori:** Single beds. They're—
4. **Sam:** Y'mean narrow?
5. **Lori:** They're awfully narrow yeah.

In line 1, "thin" seems to be an unusual choice of word for this context. It is unclear what Lori means. "Thin" might be more normally applied to a different dimension of a bed, namely how thick a mattress is, not how wide the bed frame is. Sam's first attempt at initiating repair uses the general term "What?" Lori starts to clarify this in line 3 when Sam re-does the initiation of repair, this time in much more specific form, no longer requiring explanation from Lori but simple confirmation. Again, the pattern is to upgrade by being more specific in the form of repair initiation.

Psychologists Herbert Clark and Edward Schaefer examined this phenomenon in a collection of more than 750 telephone calls to a directory enquiries number in England. In these calls, there was very often some back-and-forth between the caller and the operator about whether a requested number was clearly heard or accurately recorded. Here is an example:[8]

1. **Operator:** Directory Enquiries, for which town, please?
2. **Caller:** In Cambridge.
3. **Operator:** What's the name of the people?
4. **Caller:** It's the Shanghai Restaurant, it's not in my directory, but I know it exists.
5. **Operator:** It's Cambridge 12345.
6. **Caller:** 12345.
7. **Operator:** That's right.
8. **Caller:** Thank you very much.
9. **Operator:** Thank you, good bye.

The important lines are 5–7. The operator gives the crucial information being requested—the number of the restaurant—and the caller checks to confirm that they have heard this correctly. This checking move in line 6 operates like the strong other-initiated repair strategies we have seen above (compare the "till Monday" and "single bed" examples), in that the resolution of the problem is done with a simple confirmation ("Yah," "Yeah," "That's right").

Now imagine a different version of the same interaction, in which a more general repair initiator is selected:[9]

5. **Operator:** It's Cambridge 12345.
6. **Caller:** Pardon?
7. **Operator:** It's Cambridge 12345.
8. **Caller:** Thank you very much.

Here, if a caller were to use the much weaker repair initiator "Pardon?," it would still lead to resolution of the problem. The end result would, in a sense, be effectively the same. The caller would receive confirmation in the next turn (line 7) that they had correctly heard the number, and then the conversation would move forward. But there is an important difference between the two versions. The difference has to do with the amount of work that the operator is required to do in line 7. In the first version, all the operator had to do was confirm: "That's right." In the second version, they have to repeat the entire prior turn.

Looking at their collection of phone calls, Clark and Schaefer found that callers hardly ever used the weak strategy, relative to the stronger or more specific strategy. In only 4 percent of the cases did the caller use a weak type of repair initiator such as "What?" Meanwhile, more than two-thirds of the calls featured the very strong strategy of repeating the entirety of what was heard, meaning that the operator only had to give a simple confirmation. Callers consistently opted for the strongest form of repair initiation that they could manage. This minimized the trouble that the operator would have to go to in order to fix the problem. On the basis of these observations, Clark and Schaefer proposed that when people initiate repair on others' turns in conversation, they follow a "strongest initiator rule."[10]

It is worth asking whether the tendency to be as strong or specific as possible when initiating repair is specific to contexts such as formal interactions with telephone operators. Our twelve-language cross-cultural study of other-initiated repair provides a way to test this, as we focused on everyday interaction in informal settings, between people who know each other well: relatives, friends, neighbors. We collected examples of other-initiated repair from free-flowing conversations in home and village settings, and we checked to see if people similarly opted for more specific ways of initiating repair.

If people generally prefer to use more-specific available forms of repair initiation, then this implies that when people use an especially weak form, such as "What?" or "Huh?," it is because they had no choice. They could not have been more specific if they had wanted to. They could literally do no better than "What?" or "Huh?" because, for example, they did not hear any of what the other person just said, their attention was elsewhere, or the thing that was said was completely unexpected.

We examined the cases of other-initiated repair in our twelve-language sample and identified the instances in which the following three common causes of trouble with hearing or understanding applied. First, there was obvious noise interference (for example, somebody else also speaking at the time or a door being slammed). Second, Person B's attention was elsewhere (for example, Person B was talking to somebody else or looking at their phone). Third, Person A's turn was a question (being a question, it is more likely to be something unrelated to everything said so far, and therefore not anticipated by the listener).

We isolated the examples in our data in which all three of these things were true at the same time. We called these trouble-prone contexts. In trouble-prone contexts, multiple factors conspired to create the worst possible conditions for hearing or understanding what a person just said. We found that if a repair initiator was produced in this type of context, we can predict with 99 percent certainty that it will be of the weak or nonspecific type: that is, a form such as "Huh?" or "What?"

It is of course not surprising that when people truly cannot be specific, they will not be specific. In a trouble-prone situation, a person usually has no choice but to be vague when initiating repair. But in other contexts, people do have a choice. In the Clark and Schaefer study of directory inquirics in Cambridge, we saw that people follow a strong initiator rule. They are very specific when checking on their understanding of what was just said. To see if people behave like this in regular face-to-face conversation, we looked at the examples in our cross-linguistic study in which *none* of the three elements of trouble-prone contexts applied. We labeled these the default cases. We then looked at how frequently the more specific repair initiators are used in default cases. We found that if a repair initiator is used in a default case, in which the conditions for hearing and understanding are favorable, around two-thirds of these will be of the strong, specific type such as "Who?"

Notice that in these cases, whenever a person has been more specific, for example saying "Who?" as in the "Sibbie's sister" example, they could always have chosen to be vaguer than that (e.g., saying "Huh?" instead). It is not a given that people will be more specific in initiating repair, so it is reasonable to ask whether people who speak languages other than English do things differently.

The many studies of repair in English imply that their findings tell us about human interaction generally, but they do not make explicit claims about universal implications of their findings. They leave untested the question of whether the principles found are independent of the language or culture of the people involved: English versus any other of the world's 6,000 or so languages. Our twelve-language study was designed to address this. With data from languages around the world, we wanted to test whether people everywhere follow the same rules in other-initiation of repair.

The first thing we did was to check whether people everywhere carry out the repair behavior using the same sequence pattern that is found in English. We found that in all of the languages we sampled, these sequences happen all the time. They have the same structure that we have seen in the English examples, where a side sequence intrudes into the progression of the conversation to solve a problem before resumption can occur.[11] The structure is illustrated here, with annotations on the familiar Dippert example:

TROUBLE SOURCE	**A:** Dippert's there too.
REPAIR INITIATOR	**B:** Huh?
REPAIR SOLUTION	**A:** Dippert is there too.
RESUMPTION	**B:** Oh is he?

Not only did we find that these sequences occur in all the languages; we also found that they are abundantly common. In a total of 4.5 hours of running conversation sampled, we found that on average once every 84 seconds somebody halts the conversational proceedings to correct a problem. We found that 95 percent of independent repair initiations happen within about four minutes of the last one. These findings are summarized in Figure 7.2. As time passes after the last occasion on which somebody used a repair initiator (shown along the *x* axis), the likelihood of another repair initiator being used increases (along the *y* axis).

This figure shows the pattern observed across all of our examples from all of our sample languages put together. When each language is separated out, the basic pattern is the same, though there are minor differences between the languages. For example, in our Icelandic examples, repair was initiated at a slightly lower rate than in Argentine Sign Language.[12]

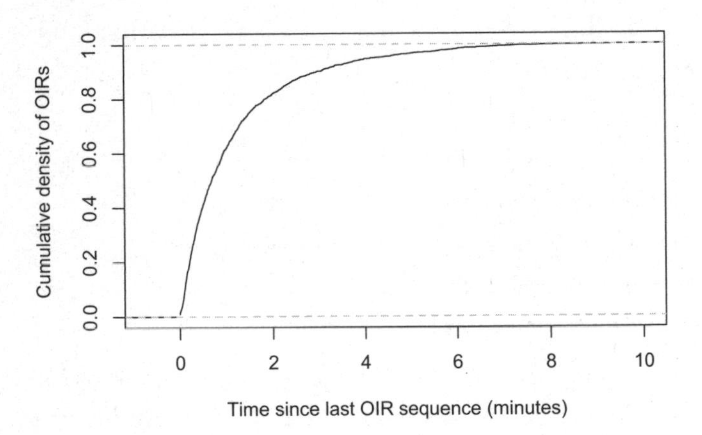

Figure 7.2 Likelihood over time that an independent sequence of other-initiation of repair (OIR) will occur (all languages together). Data from Dingemanse et al. 2015.

These data show the overall frequency of repair initiation, without distinguishing between the different types. We checked if all the languages we were studying provided people with the same distinctions between weak repair initiators (such as "What?" and "Huh?," typically requiring a full repetition from Person A) and strong repair initiators (of the two subtypes we have discussed: those that ask for a specific piece of information and those that require a simple confirmation). The answer was a clear yes.[13] In any language we looked at, the same basic options for repair initiation are available. But there are interesting differences in the details.

One of the clearest differences between languages has to do with the inventories of weak repair initiators. Languages typically provide numerous types of weak repair initiator. In English, there are three broad categories. The first is the "Huh?" type, which includes variants such as "Hmm?"[14] This type uses a simple word that is dedicated to the repair initiation function. This type of word is called an interjection, a kind of word that can make up a complete utterance in itself and that is not normally used in complex phrases.[15] Other examples of interjections in English include "Wow!," "Yuck," and "Phew."

A second type of weak repair initiator is the "What?" type (with subtle variants in English such as "Wha?"). This type uses a word that also has broader functions in the language, for example occurring in other sentences as a regular question word. In English, "What?" can be used by itself as a repair initiator, and it also can occur as a word within a full sentence such as "What flavor is that?" or "They heard what you said."

A third type of weak repair initiator is the formulaic type. These are special idioms for initiating repair. In English, this last type includes the so-called polite options that parents teach their children to use. These include "Sorry?," "Excuse me?," and "Pardon me?"

Our study found that while all languages have weak repair initiators, they do not necessarily have all three of the subtypes. The only subtype that appears to be truly universal is the interjection "Huh?" This is the topic of the next chapter.

The second type, using a question word such as "What?," is very widespread. Most languages have something like it. Here are three examples:

Cha'palaa (spoken in Ecuador)[16]

A: pikishnetyuu mama
Don't make the floor vibrate, daughter.

B: tin
What?

A: pikishnetyuu tya'pumi
Don't make the floor vibrate, (the camera could) fall.

Siwu (spoken in Ghana)[17]

Mum: Sesi su ɛɛ̀ iraɔ̀ tã mɛ
Sesi take uh the thingy for me

Sesi: be

What?

Mum: su kadadìsɛ̃ĩbi bɔ mɛ

Take the small pot and bring it to me.

Italian[18]

Eva: quel coso lì èl quel del gat Mario uh Mirko

That thingy there is it for the cat, Mario, uh, Mirko?

Mirko: cosa

What?

Eva: quel piatim lì

That saucer there

Mirko: sì

Yes

Most languages in our study use a question word for repair initiation in this way, although there are exceptions. In two languages in our sample, there were no examples of "what" being used for other-initiation of repair: Yélî Dnye and Tzeltal.

Formulaic strategies of the "Sorry?" and "Pardon me?" variety turn out to be rare in the world's languages. English is in a minority among languages of the world in this respect. Many languages do not have repair initiators like this at all. Polite forms of speech tend to be well developed in languages that are spoken in large-scale societies, where language is often used in institutional interactions, and for speaking with strangers. In smaller-scale and more rural societies, what we think of as polite forms of speech may play little or no role.

Despite some differences between languages in the finer details, the basic principles of systems for repair initiation are the same around the world. In each language, there is a three-way distinction in types of other-initiator of repair, going from weaker to stronger. No matter the language spoken, when a person wants to initiate repair, they must choose from among these three options.

In our research group, we wanted to find out whether speakers of all languages followed the same principles in selecting from among these alternatives in conversation or whether there is cultural variation. One possibility is that people simply follow the same rational principles everywhere. Or it may be that different strategies reflect different cultural values. Some researchers have proposed that people in certain cultures show a preference for leaving things vague.[19] It is perfectly conceivable that in certain cultures, such a preference would lead to the weak form of repair initiation being the most dominant type.

For every case included in our study, we noted whether or not the person who initiated repair had their attention on something or someone else when the trouble turn was spoken. We found that when the person was involved in a parallel course of action, they would use a weak form of repair initiator about half the time. But when their attention was on the speaker, they were twice as likely to use a strong initiator, producing weak initiators only a quarter of the time. The same pattern was observed in all the languages.

A second issue concerns whether the trouble turn was less expectable. Consider the following case, from a study by the sociologist Paul Drew:[20]

1. **Gordon:** Hi Norm.
2. **Norm:** Hi Gordy.
3. **Gordon:** Um are you going tonight?
4. **Norm:** Mm.
5. **Gordon:** Would you mind givin me a lift.
6. **Norm:** No, that's all right.
7. **Gordon:** Very kind of you.
8. **Norm:** Caught me in the bath again.
9. **Gordon:** Pardon?
10. **Norm:** Heh. Caught me in the bath.
11. **Gordon:** Oh I'm sorry well I should let you get back to it.

Look at the highlighted lines, 8 and 9. Drew points out that there is a recurring pattern involving utterances like those in line 8. He noticed two things. First, these utterances make a sharp deviation in topic from the conversation up until that point. Second, they result in a weak repair initiator: "Huh?" or equivalent. This led us to wonder whether moves that are harder to predict might therefore be harder to process, and in turn might be more likely to elicit a weaker form of repair initiator. To study this in a systematic way, we focused on two kinds of utterance that contrast greatly in terms of how well we might anticipate them in conversation.

If someone asks "What time is it?," we can predict with some certainty the kind of answer this question will receive. Theoretically, the possible answers are infinite. But in practice we can be fairly sure that respondents will do one of a small number of things. They could state a time of day using a standard format ("It's half past two," "Five to eight," "Four-thirty"). They could refer to the time in some other, indirect way ("Time to leave," "Almost lunchtime," "My parents won't be back for another hour"). Or they could give a reason why they can't answer the question ("I don't know," "I don't have a watch"). If utterances that come immediately after questions are easier to anticipate than questions themselves, then this implies that responses to questions should cause fewer problems with processing or understanding.

We checked to see if this is the case, and we found that when the trouble source was a question, nearly half of the repair initiators used were of the weak type. But when it was an *answer* to a question, the likelihood of people using a weak repair initiator was cut in half, with the weak type used only a quarter of the time. This pattern held up across all twelve languages in our study.

A final point of cross-language comparison for relative usage of weak and strong repair initiators was in sequences involving repeated attempts at repair. It happens quite frequently that a first attempt at repair does not lead to a satisfactory resolution, and another round is needed. Here is an example:[21]

1. **Mum:** Oh what is it.
2. **Leslie:** What?
3. **Mum:** The place.
4. **Leslie:** What place?
5. **Mum:** I asked you in my letter.
6. Where—Where um in—
   Waithe's garage is.
7. **Leslie:** Sparkford.

In this example, the first attempt at initiating repair is of the weak type (in line 2), and it is upgraded to a stronger type for the follow-up attempt (in line 4). We looked to see if the same pattern is found in other languages. We identified all of the repair initiators that occurred as a follow-up to a failed attempt at initiating repair. We found that when a repair initiator is first in a sequence, about a third of the time the weak type will be used. But when it is a follow-up, the proportion of weak types used drops to half that amount. This pattern shows that when people get a second chance to initiate repair, they will be more specific. This is evidence that the strong initiator rule proposed by earlier authors is not just a rule for English but also a principle behind the organization of repair in all languages.

Taken together, the findings we have discussed in this chapter reveal a universal tendency for people to be as specific as they can when initiating repair on others' speech in conversation. This tendency tells us something important about the cooperative character of conversation. Being more specific in initiating repair has two advantages. The first is that it increases the chances of the problem being solved in one go, allowing the conversation to get back on track. This benefits both parties in the conversation, and as such it is a cooperative strategy.

The second advantage of being more specific reveals a possibly more altruistic motive.

A: Dippert's there too.
Scenario 1 **B:** Huh? **A:** Dippert is there too.
Scenario 2 **B:** Who's there too? **A:** Dippert.
Scenario 3 **B:** Dippert's there too? **A:** Yes.

Compare the three scenarios in terms of the division of labor between Person B and Person A in resolving the problem. The three scenarios differ in the strength of the repair initiator used, from weakest to strongest. Now compare how much work each person has to do in these scenarios, relative to the other person. As the repair initiators get more specific, Person B has to do more work, while Person A has to do less. When Person B chooses to be more specific, they are choosing to put more work into solving the problem, and at the same time they are requiring the other person to do less. A full repetition—"Dippert is there too"—might not take a lot of effort, but it certainly takes a good deal more effort than a simple "Yes."

We measured the relative "cost paid" by the two people involved in all of our examples, and we found that the tendency holds across the languages.[22] The result shows that regardless of the language being spoken, or the culture in which it is spoken, people opt to be more specific in initiating repair, and by doing so they reduce the amount of effort the other person has to go to in order to fix the problem.

These principles for repair emerge in the context of a universal conversation machine. Speakers of all languages use the same basic options and select them according to the same principles. The system is grounded in a shared willingness to cooperate, and thereby to maximize efficiency in interaction. When using the system of other-initiated repair, people are not only able to make conversation more efficient; they do so in a prosocial way, by minimizing the effort that others have to go to in repairing problems as they arise. The system of repair keeps conversation running smoothly by ensuring that people are in sync at each step.

We have focused in this chapter on broad principles of repair, based on an unprecedented sample of languages from across the world. In doing our comparative research, a feature of the system of repair stood out. In one specific strategy for other-initiation of repair, languages showed much closer convergence than would be expected. We now turn to the discovery that the word "Huh?" appears to be a human universal.

8
THE UNIVERSAL WORD: "HUH?"

In a village of the riverine plains of central Laos, a man calls out through thin bamboo walling to a next-door neighbor named Noi: *Noi bòò mii sùak vaa Noi* ("Noi, do you have any rope Noi?") She calls back: *Haa?* ("Huh?") The man repeats the question: *Bòò mii sùak vaa* ("Do you have any rope?")

A hillside hamlet in central Eastern Ghana. Two speakers of the minority language Siwu are preparing gunpowder, which will be sold for use at a funeral in another village. One man asks: *Ilɛ̀ isɛ̀-ɛ?* ("Where is the funeral?") The other: *Hã?* ("Huh?") The first: *Ilɛ̀ isɛ̀-ɛ?* ("Where is the funeral?")

In the coastal lowlands of northwestern Ecuador, two Cha'palaa speakers exchange information. *Motorkaa detisaa* ("They say he bought a motor.") *Aa?* ("Huh?") *Motorkaa detisaa* ("They say he bought a motor.")[1]

This same sequence of other-initiated repair is found in daily conversation in every household, village, town, and city of the world, regardless of the culture and lifestyle of the people or the language they speak. But in these three examples something more is going on. Not only does each of the languages have a way of doing the weak type of repair initiation, but the word used for doing it sounds almost the same in three corners of the world. It sounds very much like the English word "Huh?"

Members of our research team initially saw the possibility that "Huh?" was a universal word in the context of the large cross-language project on repair in conversation discussed in the last chapter. We were focused on how people deal with misunderstandings in language. In a preliminary phase of group research, we were struck by a repeated observation. In all the languages for which we had data, people were saying something that sounds pretty much like English "Huh?" and has pretty much the same function. This led us to ask: Could "Huh?" be a universal word?

To really know for sure whether "Huh?" was universal, we would have to check every single one of the 6,000 or so languages spoken in the world today. This is of course also true for any of the proposed universals of language that you'll come across in linguistic research. The linguist Anna Wierzbicka argues that a small set of around 60 word meanings are universal. Every language, she says, has a clear way to express simple concepts including "good," "all," "people," and "you." The linguist R. M. W. Dixon argues that every language has adjectives, a class of words that are distinct from nouns and verbs. And the linguist Noam Chomsky argues that recursion—taking the output of a process and using it as input for the same process again—is found in the grammatical structure of all languages.[2]

When these linguists give evidence for their claims, they can cite only a small subset of the world's languages. But they are still able to state that something is highly likely to be universal, as long as no counterexamples have so far been found. This is because linguistics is an inductive science. As linguists gather information about what is possible in the different languages of the world, they can become more and more certain about what they will find, or not find, in languages they haven't looked at yet. That said, they must always be open to finding new information that disproves their claims.[3]

Table 8.1 shows weak repair initiators in 15 languages.[4]

All of the words in both columns 1 and 2 can be used for the same general function as "Huh?" The words in Column 1 are all questioning words that mostly mean "what"—or sometimes words that mean "how" can be used for this function, as in the optional forms in Italian

Language	Where	Column 1	Column 2
ǂĀkhoe Hai‖om	Namibia	mati	hɛ
Cha'palaa	Ecuador	ti	a:
Chintang	Nepal	tʰɛm	hã
Duna	PNG	aki	ɛ̃:
Dutch	Netherlands	wat	hɜ
English	UK	what	hã:
French	France	quoi	ɛ̃
Hungarian	Hungary	mi	ha
Icelandic	Iceland	hvað	ha:
Italian	Italy	cosa	ɛ:
Lao	Laos	iñaŋ	hã:
Murrinhpatha	Australia	t̪aŋgu	a:
Russian	Russia	shto	ha:
Siwu	Ghana	be:	hã
Spanish	Spain	que	e

Table 8.1 Weak repair initiators in 15 languages, showing both the "question word" forms (in Column 1) and the "interjection" forms (in Column 2). The words in Column 2 (and some in Column 1) are written using the International Phonetic Alphabet, offering a direct representation of the sound of these words.

(*Come?*), German (*Wie?*), and French (*Comment?*). You will also notice that these words sound completely different across languages: compare *what* in English, *shto* in Russian, *aki* in Duna (a language of highland Papua New Guinea). This is what we expect to find when we compare words from different languages: compare the words for "dog" in the same three languages: English (*dog*), Russian (*sobaka*), Duna (*yawi*).

Now look at Column 2. Something different is going on. There is an uncanny resemblance between these words. All spoken languages use vowel sounds. They each draw on the same human set of possibilities. In Figure 8.1 the possibilities for vowel sounds in human languages are laid out on a grid, with two axes: front-to-back and close-to-open. Each phonetic symbol on the chart represents a vowel sound that is known to be used in at least one human language.

The "what" words in Column 1 of Table 8.1 use vowel sounds from all over this chart. But our observation of 20 or so languages showed that in the expressions corresponding to "Huh?," not a single language used a sound

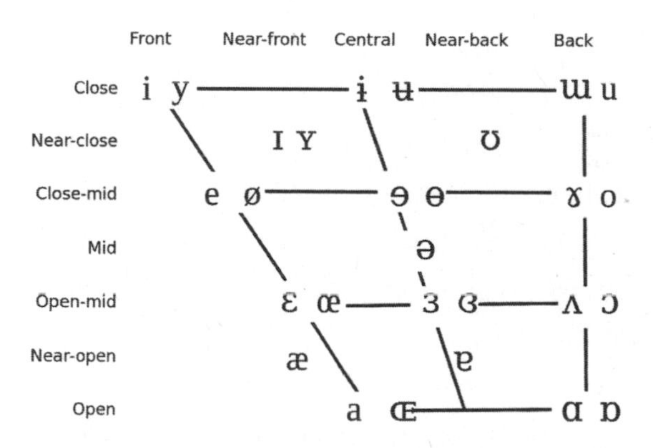

Figure 8.1 International Phonetic Alphabet symbols for vowel sounds that are used in human languages, arranged in a grid according to the position of the vocal apparatus. Source: https://www.internationalphoneticassociation.org/content/full-ipa-chart.

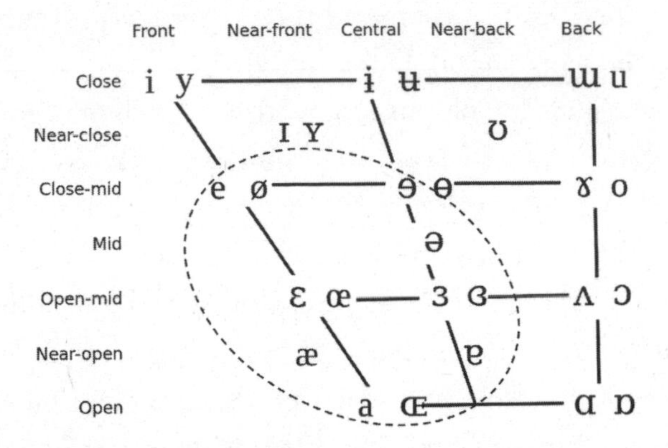

Figure 8.2 International Phonetic Alphabet symbols for vowel sounds: circled area shows sounds that occur in words meaning "Huh?" across 31 languages. Source: https://www.internationalphoneticassociation.org/content/full-ipa-chart.

from outside the open-front quadrant of the space, circled in Figure 8.2.

Together with linguists Mark Dingemanse and Francisco Torreira, I wanted to find out just how similar in sound all these "Huh?" words actually are. We isolated examples of the single word "Huh?" supplied from recordings of everyday conversation in ten languages, with the goal of plotting and objectively comparing the sounds of them.

To get to the essence of what we found, let us compare two of the languages: Spanish and Cha'palaa.[5] All of the "Huh?" words from Spanish and Cha'palaa have a vowel that is located just within the open-front quadrant circled in Figure 8.2. This quadrant includes vowels like those found in the English words "bed," "bad," and

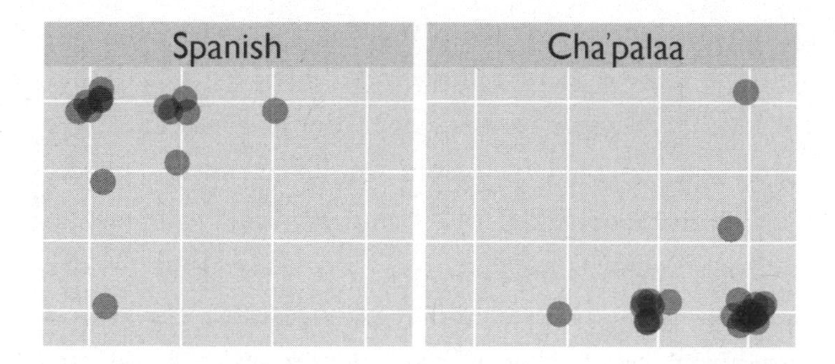

Figures 8.3a and 8.3b Comparison of the vowel sound used in instances of "Huh?" from recorded conversations in Spanish and Cha'palaa. Each square represents the same "open-front" region of vowel space, with each black dot corresponding to the measured location in the sound space of each recorded instance of "Huh?" Adapted from Dingemanse, Torreira, and Enfield 2013:4.

"bud," and does not include vowels like those in "bead," "booed," and "board." That makes the languages surprisingly similar. But when we zoom in on this quadrant and compare languages, we find that they differ in clear ways.

Figure 8.3 zooms in on the bottom-left quadrant circled in Figure 8.2. Each dot in Figure 8.3a and Figure 8.3b represents the vowel used in each instance of "Huh?" in recorded Spanish and Cha'palaa conversations.

Figure 8.3a shows that the Spanish versions of "Huh?" are mostly grouped in the small area of the quadrant corresponding to the sound "e," rhyming with the vowel in the English word "bed." Figure 8.3b shows that the Cha'palaa versions of "Huh?" tend to land on a different spot, corresponding to the sound "a," similar to the vowel heard in the English word "cat." These are different-sounding vowels, different enough that English can use this distinction alone to tell between a "fed" and a "fad," or a "wreck" and a "rack." Some researchers have proposed that interjections like "Huh?" are not real words, but are grunts of some kind.[6] But if "Huh?" were just an animal-like signal of surprise, then it should not differ systematically in these (albeit minor) ways. The vowels used in the Spanish and Cha'palaa versions of "Huh?" are different enough from each other that Cha'palaa children must be learning to say the word differently from the way that Spanish children do. The local version of "Huh?" is specifically shaped to the language. It is a real word, not a grunt. It is incorporated into the local language system.

To say that this word varies in form depending on the language system it is embedded in is not inconsistent with saying that it is universal. All humans are capable of producing the same wide range of vowel sounds. In Figure 8.3 we see that "Huh?" always occurs with a vowel in the open-front corner of the vowel possibility space. Figure 8.4 makes the same point, this time with each of the languages from our ten-language study plotted at the average point where all of the examples of "Huh?" from that language landed.[7]

There is a remarkable convergence across the languages here. While there is indeed some variation in the precise point at which this word is pronounced in each language (as just described for Cha'palaa and Spanish), the vowel of the "Huh?" word in these languages shows

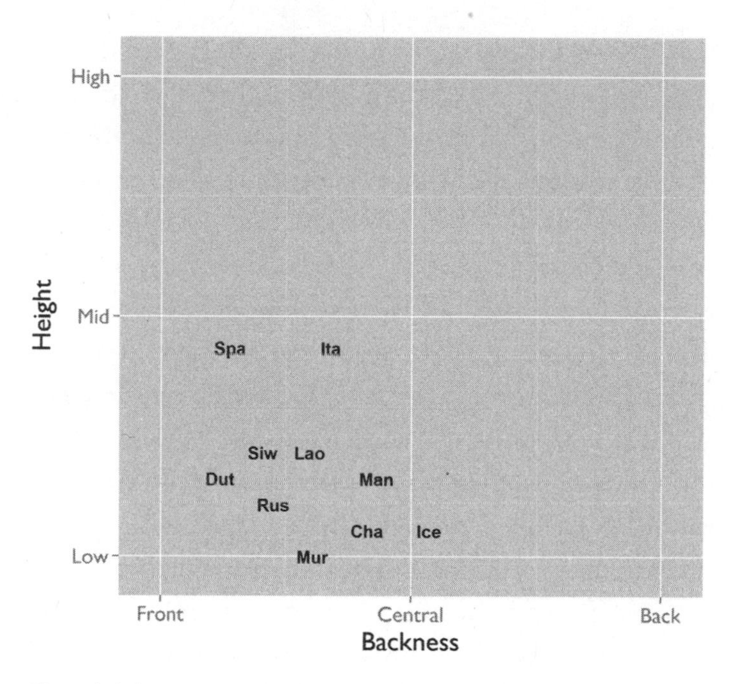

Figure 8.4 Average positions of the interjections in vowel space. The vowel inventories of the world's languages tend to make maximal use of vowel space. In contrast to this, the vowels of the "Huh?" interjections all cluster in the same lower-front region. Abbreviations: Cha, Cha'palaa; Dut, Dutch; Ice, Icelandic; Ita, Italian; Lao (Lao); Man, Mandarin; Mur, Murrinhpatha; Rus, Russian; Siw, Siwu; Spa, Spanish. Source: Dingemanse, Torreira, and Enfield 2013: 4.

much less variation than would normally be expected for a random word across languages. It never lands in a high or back area of the vowel space. It is always found in the lower-front corner of the space.

When we say that "Huh?" is universal, we mean that in all the languages we've investigated or heard about so far, there is a word that sounds like "Huh?" and that means the same thing. We can be proven wrong. But we are confident that the word is universal because the languages we sampled come from a broad range of different language families.[8] It would have been no good checking for a universal on the basis of languages from the same language family (e.g., Russian, Italian, German, French, Spanish, and English—all Indo-European languages). This is because those languages have a single common ancestor: Maybe the "universal" feature in question was a rare innovation that occurred once in history, in just that ancestor language, and all the modern descendants simply inherited it from the same source. Nineteenth-century polymath Sir Francis Galton pointed out this problem in his commentary on the work of the anthropologist Edward Tylor, insisting on the need to ensure that the individuals in a sample are independent.[9] In linguistics, this means that if we are going to claim that something is universal, we had better test languages in a sample from as many language families as possible.

In our study, we had data from 31 languages in total, and these derived from 16 distinct language families. This sample is only a fraction of the world's languages, but in practice, a sample of this size and kind is sufficient to falsify many possible claimed universals. If "Huh?" were not universal in the sense we claim, chances are high that at least one of the 31 languages would lack it.[10]

The discovery of a universal word is not what received wisdom in linguistics would predict. A century ago, the Swiss linguist and semiotician Ferdinand de

Saussure established the doctrine of the arbitrariness of the sign.[11] His example of a sign was the word "tree." The sign has two parts. One part is a sound. In his example it is the sound we make when we pronounce the word "tree." The other part is a concept, something like what we think of when we hear someone say "tree." The arbitrariness of the sign refers to the link between a sound and a concept. Saussure said that this link was entirely arbitrary. It is the result of historical accident in a language. There is no significance to the fact that the English word "tree" has the sound that it does. Basically, the same concept is connected with completely different sounds in other languages, by sheer convention: *baum* in German, *ton-mai* in Thai, *kaspi* in Amazonian Quechua, and so on.

Obvious exceptions to the rule are imitative or onomatopoeic words, including "plop," "quack quack," and "cockadoodledoo." But these are not as similar across languages as you might expect. Speakers of the Lao language say *djoum*, *gaap gaap*, and *oki ok ok* for these meanings. The sounds of these words are arbitrary in the sense that they can be quite different from language to language. But they are motivated. When you learn the word, you can immediately see a sensible connection between the sound and the meaning. "Quack quack" sounds something like what a duck says. And so does *gaap gaap*.

Then there are more subtle cases of sound linking to meaning in motivated ways. Here are two more examples from Lao. If you *ah* your mouth, it means that you open it up wide ("Say *Ah*!"). And if you *mim* your mouth, it means you press the lips tightly together. The sounds of these words connect to the words' meanings through the actions that are required for pronouncing them. For reasons like these, unrelated languages sometimes have similar-sounding words for similar meanings. If a word is uncannily similar across unrelated languages, it may be because there is a common motivation for languages to independently evolve such a form.

Along these lines, we argue that "Huh?" is the product of convergent evolution. This term refers to the independent evolution of similar structures in unrelated species. For example, dolphins (a species of mammal) have independently evolved very similar body shapes to sharks (a species of fish), and they have independently evolved similar systems of echolocation to those used by bats.[12] Convergent evolution occurs when two species respond in similar ways to similar environmental pressures.

"Huh?" has a similar form across languages because the same set of conditions leads in all languages to something like the "Huh?" word being produced. In the flow of conversation, people need to be sure that others know when they have failed to understand. As we learned in earlier chapters, time runs by quickly in conversation, and one's chance to signal that there is a problem can pass quickly. In that situation, you need a syllable that is quick and easy to pronounce. "Huh?" does the job. The particular vowels that "Huh?" is restricted to in all languages happen to be the vowels that are most easily pronounced when a person's tongue is in a relaxed position.

Think about the conditions a person is under in conversation. First, not only are people capable of responding to what others say within the time it takes a sprinter to react to the starter's gun, but there is a social expectation that they will do so. Second, there is a social preference not to hold up the conversation by requiring others to go back and revise. It is better to avoid initiating repair. Third, there is a further preference—which competes with the last one—for ensuring that problems in common understanding do not pass by unresolved.

Ideally, in conversation a person hears and understands everything the other person said, so these preferences do not come into conflict. But when a person finds that they have failed to hear or understand what someone just said, there is a dilemma. Time is passing quickly. To avoid halting proceedings, a person might first try hard to process or reconstruct what they have heard. But with every passing millisecond, any delay in response increases the risk of adverse consequences. One risk is that, as we saw in Chapter 4, a delay can be interpreted as a negative response to what was just said rather than simply a sign of inability to respond.

Once a delay becomes too long, and especially when it begins to extend beyond the 600 millisecond mark, creeping into the late zone of the one-second window for turn transition, a speaker needs to act fast. The thinking process goes something like this: "I've tried to process this but have failed to do so in the time I had; now it's getting late, and the other person will either keep going, unaware that I haven't understood, or they will figure I'm unhappy with what they've said; now I need to make a sign quickly." By this point, the speaker needs to launch a recognizable sign as fast as humanly possible and with minimum effort. As we learned earlier in the book, it takes a good 600 milliseconds to go through the mental processes of finding a specific word and instigating a specific motor program for articulating that word. The advantage of the sound of "Huh?" is that it requires a minimum of motor planning and execution. With the tongue in a neutral and relaxed state, and with the mouth slightly open, a person needs only to make a vocal sound with some force, and the sound that will come out is "Huh?"

The fact that "Huh?" can be uttered with minimal delay does not mean that the word will be faster than other kinds of responses. Research on English shows that "Huh?" is heard on average after a delay of 835 milliseconds, much later than the average turn transition of around 200 milliseconds (and indeed later than strong repair initiators, which come after an average pause of 726 milliseconds).[13] One reason why "Huh?" is both fast to produce and late to arrive is that the speaker will often have used up time trying to process or reconstruct what they just heard. Upon failing to do so, even a very fast production time will still leave them in the late zone of the one-second window. Better late than never!

Our claim about convergent evolution is that the same solution to the problem arose in all languages because all languages share a common environment: a high-speed forward-feeding system of turn transition in which participants are accountable for keeping up. "Huh?" sounds the same across languages because that particular sound is best fitted to the word's universal function: the need to signal as quickly and simply as possible that there is a general problem.

When we reported our finding that "Huh?" is a universal word in human languages, people liked the irony of it: "One of the few things that cultures share is a sign of misunderstanding and confusion."[14] But "Huh?" does not stand for universal confusion. It stands for universal cooperation. It shows that there is a global need, and willingness, to pause a conversation and sort out a communication problem as it occurs.

This little word, like the turn-taking system it operates in, suggests a moral architecture to communication. The word makes sense only when you can assume the other person will cooperate, by backing up and repeating what they said, thus helping you stay on track. This is possible because conversation involves joint commitment, and the interpersonal accountability that goes with it. And it explains why no other species goes "Huh?" or does anything like it, even when their communication systems are complex. If the complexity of the signal were the key point, it would be surprising that no other animal goes "Huh?" The word has none of the complex grammatical structure that defines human language. In fact, it has no grammatical structure at all. But this is not what makes "Huh?" possible. It is possible only when an ultra-flexible system like language, with ultra-cooperative users, demands and supplies a solid and readily available backup mechanism for ensuring intersubjectivity at every step.

9
CONCLUSION
THE CAPACITY FOR LANGUAGE

We are on the verge of a full-blown scientific revolution in research on the human capacity for language.[1] To complete this revolution, a new generation of language researchers must build a bridge across the chasm that separates two very different ideas. On one side is the idea, captured in our notion of a conversation machine, that there is no capacity for language without specifically human forms of social cognition and interaction.[2] On the other is the idea that language is a private, purpose-specific computational system for operating upon information.

These ideas about the essence of language point to a long-standing debate about the human capacity for language. Humans have a natural capacity, which no other species has, that makes it possible for infants to learn any language they are exposed to, within a few short years and with no formal instruction. Scientists of language describe this capacity in two senses, which we can call broad and narrow:

1. *The capacity for language (in a broad sense)*
   Humans have something that makes language possible in only our species.
2. *The capacity for language (in a narrow sense)*
   Humans have something that makes language possible in only our species, and that thing is innate linguistic knowledge.

Sense (1) is an axiom. We know the statement is true, and it leaves open for research to determine just what sort of thing the capacity is. Sense (2) is a specific hypothesis, sometimes referred to as *Universal Grammar*.[3]

Noam Chomsky and colleagues have long proposed that people have innate knowledge of linguistic principles. For example, a principle of *structure dependence* states that rules of grammar are defined with reference to grammatical phrases, not words.[4] The rule that allows us to take "John ran away" and rephrase it as "Away ran John" might look like a reversal of the order of words, but the rule cannot be described as "Reverse the order of words." We know this because it is impossible to take

"The boy ran away" and rephrase it as "Away ran boy the." The rule must refer to higher levels of structure such as grammatical phrases (e.g., "the boy").

Another example is a principle of *subjacency,* which puts limits on which bits of a sentence can be moved around, and to where. If I say "He bought a green shirt," you can form a question from that by replacing the whole phrase "a green shirt" with "What" and moving it to the start of the sentence: "What did he buy?" The subjacency constraint stops you from moving just a *part* of that phrase. English rules won't let you say "What did he buy a green?"

A final example is *recursion,* when the output of a procedure is used as input for running the same procedure again, producing new structures of theoretically infinite length.[5] The sentence "He knew it" can be expanded by inserting a sentence, "Mary was a lawyer," in place of "it," giving us "He knew Mary was a lawyer." In turn, this sentence can be embedded in a new sentence: "Bill said he knew Mary was a lawyer." Then, "Kim thought Bill said he knew Mary was a lawyer." And so on, ad infinitum.[6]

Formal principles like the three just mentioned are often put forward as evidence that the language capacity must consist of inborn knowledge that is specifically linguistic in nature. But many linguists take a different view, arguing that there is no inborn knowledge of language. Many believe that our knowledge of language is built up entirely from experience and that principles such as structure dependence, subjacency, and recursion emerge from psychological capacities that are not specific to language. These capacities have cognitive and behavioral functions beyond the learning and processing of language, for example general statistical learning, imagistic thinking, and metaphor.[7] Linguists including Adele Goldberg, Joan Bybee, and Ewa Dąbrowska, among many others, have used these ideas to explain how people can learn and use language without having to rely on innate linguistic knowledge.[8] In this view, the human capacity for language is like the human capacity for chess. Humans are the only animals that play chess, but we are not born with evolved chess-specific knowledge.[9]

For the last 30 years, linguists have quarreled over which of these views of the language capacity is correct.[10] The debate has generated a number of breakthroughs in our understanding of language. But it has also reinforced a major blind spot in the mainstream of research because both sides of the debate use similarly narrow definitions of the two key concepts, *language* and *cognition*.[11] By "language" they mostly mean the structural properties of words, phrases, and sentences. By "cognition" they mostly mean the abstract structures in the mind that underlie words, phrases, and sentences. They have focused on linguistic structures *inside* phrases, but they have seldom looked at the relations *between* phrases as turns in conversation and in larger stretches of discourse.[12] And they have looked at how information is represented in the minds of individuals, but not at the role of social cognition in the cooperative use of language.

To see what is being bracketed out from all of this linguistic work, let us take a final look at the familiar Dippert example (here with transition times from the original recording, stated in milliseconds between turns):

1. **A:** Dippert's there too.
2. (160 ms)
3. **B:** Huh?
4. (140 ms)
5. **A:** Dippert is there too.
6. (400 ms)
7. **B:** Oh is he?

Anyone with a standard core training in linguistics would look at this example and see the complex grammatical structures in lines 1, 5, and 7. They would readily describe these using technical terms such as *copula construction, adverbial complement, pronominalization,* and *auxiliary inversion.* And they would note that any theory of the human capacity for language must explain why these lines are correct in English, as well as how the English structures compare to similar structures in other languages. So far, so good. But they would have little if anything to say about line 3 ("Huh?")[13] or about the precise timings of each transition between speakers (in lines 2, 4, and 6) or about the overall structure of the sequence.

A linguist might react by saying that those aspects of conversation are not really linguistic, that they are mere performance features, representing "irrelevant conditions" of language use, as Chomsky has put it, or just general features of communication.[14] Well, for one thing, we have seen in this book that the workings of conversation are neither unstructured nor irrelevant to language. Turn-taking and repair are no less demanding of rigorous analysis and explanation than the technical structures of phrases and sentences. Another issue is that if turn-taking and repair were general principles of communication, independent from language, then we should see these same features in nonhuman communication. But we do not. Animal communication can be complex, but no animal shows the defining properties of human conversation: finely timed cooperative turn-taking, mechanisms for repair, and communicative traffic signals.

The word "Huh?" is a case in point. No other animal goes "Huh?" No other animal has a system of communication in which one party can draw attention to a problem of hearing or understanding, using the communicative system to refer to itself, and morally obliging the other to back up and repeat or rephrase. This is not because "Huh?" is intrinsically complex. It is because of the uniquely cooperative nature of language use, combined with universal rules for organizing the flow of conversation. People hold each other morally accountable to these rules. There can be a signal like "Huh?" only if players are aware of the rules and are ready to invoke them. Any explanation of the human capacity for language needs to account for this. Somewhere in the picture, the capacity for language must incorporate the socially cognitive framework of the conversation machine.

The debate about the human capacity for language has been at the center of cognitive science for decades. But self-imposed limits on the scope of research[15] have hamstrung efforts to resolve the central questions of what makes language possible in our species. The inner workings of conversation are both definitive of language and unique to our species. And, as we have seen in this book, conversation is where many remarkable and uniquely human aspects of language are found. As such, the science of language needs to explain them. But to date, we have had to look outside of linguistics to find serious attention paid to the social basis of the language capacity.[16]

Since the early 1990s, primatologist and evolutionary psychologist Robin Dunbar has worked on a theory of how language evolved in humans. He emphasizes the role of higher-order social intelligence in the context of intense social life in large groups.[17] Dunbar observes that in many social species, individuals build coalitions in order to defend one another from threats within the group or to control shared resources. Each individual must have ways of signaling their affiliations within a group—we need to know who is friends with whom. In many primate species, such as the Gelada baboons of upland Ethiopia where Dunbar has worked, this is done through personal grooming. Individuals who are socially affiliated will spend time together scratching and picking at each other's hair and skin. This behavior only incidentally clears the skin of detritus. Its real function is to create social bonds. Given that time is a finite resource, any time spent grooming one individual is time spent not grooming others. Grooming releases pleasure-inducing endorphins, and it is a public sign of interpersonal commitment between those who are grooming. When everyone in a group knows who grooms whom, then everyone knows, for instance, who will defend whom in a fight.

Dunbar proposed that language developed as a more efficient alternative to physical grooming, given that human social groups are so large. He observed that human language is mostly not used for exchanging practical information, but more for social gossip. We use language to share opinions and sentiments, and through this we establish interpersonal affiliation.

While Dunbar's work has made an impact in the field of evolutionary psychology, his ideas have mostly been ignored or dismissed in linguistics. At best, Dunbar has been chided by linguists for having a naive conception of language. The linguist James Hurford, while sympathetic with Dunbar's ideas, offers a common critique: The argument "does not say anything about the intricate grammatical structures of human languages."[18]

Michael Tomasello, also a primatologist and evolutionary psychologist, has worked to address this critique. Like Dunbar, Tomasello has developed a theory of the human language capacity that focuses on its relation to social cognition and interaction. His proposals are based not only on comparisons between humans and other species but also on experiments with young children who are learning their first language. In his book *Origins of Human Communication*, Tomasello suggests that the social functions of language are linked to specific levels of complexity in grammar. His theory recognizes three broad functions that language can have in communication: requesting goods and services, informing others about events and situations, and sharing experience and viewpoints.[19]

For carrying out the social function of *requesting* an object or service, Tomasello argues that relatively simple syntax is enough. Simple requests are primarily about objects and individuals that are present in the here and now. Requests can often be done just by pointing or simply using the word for the thing that is being requested ("water," etc.).

The next level is the social function of *informing* people of things. This function presupposes prosocial motivations—the desire to help others by giving them information. It also requires what Tomasello calls "serious syntax." When we inform people of events and situations, we need linguistic structures for referring to things that happened at different places and times, involving people and objects that are not present in the speech situation.

Finally, there is the social function of *sharing* experience, something that we do in order to build and maintain social relationships, for example by telling each other narratives and stories. This general function of language demands what Tomasello calls "fancy syntax," yet more-complex linguistic structures for keeping track of relations between people and scenes throughout a narrative. The linguistic devices needed for stories also include many of the workings of conversation, such as the ways for starting and finishing narratives in conversation that we looked at in Chapter 5. These are as important for the social function of sharing experience and opinions as the fancy bits of syntax that Tomasello alludes to.

Tomasello's theory suggests a direct link between people's social motivations in interaction and the kinds of functions that the grammar of any human language must fulfill. Tomasello goes a good way further than Dunbar does in incorporating grammatical structure in a social theory of the language capacity, but his distinction among "simple," "serious," and "fancy" syntax is still too vague to satisfy most linguists who want to explain the grammatical workings of the world's languages. To be convinced that the capacity for language is grounded in social context, linguists want to see a more explicit connection to grammar.

A clue for how to proceed can be found in new proposals by cognitive scientists—such as Morten Christiansen and Nick Chater—who forcefully reject Chomsky's claim of innate linguistic knowledge, arguing that the claim is implausible given what we know about human psychology. Following other researchers, they show that the language capacity can emerge from direct experience with language, in combination with general mechanisms of cognitive processing and principles of statistical learning.[20]

A key part of their argument is that languages are entities that evolve, like biological species. Just as each biological species has been shaped by the pressures of its environment over many generations, so has each language. The twist is in how we understand the word "environment." For Christiansen and Chater, the environment for the historical evolution of language structure is the human brain.[21] Think of it from the language's perspective. When the language is learned and used, it has to pass over and over again through the brains of its users. The language must adapt to the properties of brains: Structures that are more easily learned, processed, and remembered are more likely to evolve and survive.

But given what we have seen in this book, this can be only part of the story. The private domain of the human brain is just one environment in which language has to survive. Language scientists now need to explore the idea that language adapts not only to its repeated travels through the private domain of human brains; it also adapts to its repeated travels through the public domain of conversation.[22]

This is a distinct shift from the idea in Chomskyan linguistics that the roots of language are to be found only in underlying abstract principles. Interactional linguists including Sandy Thompson, Cecilia Ford, Barbara Fox, and Elizabeth Couper-Kuhlen have been turning that idea on its head.[23] They argue that the workings of conversation, such as turn-taking and repair, are what give grammar the shape that it has. Seán Roberts and Stephen Levinson are exploring this idea using computational modeling, to understand how grammar adapts to the pressures of the conversational turn-taking system.[24] The challenge for the next generation of researchers is to test these ideas against a broad array of differently structured languages[25] and to propose clear causal explanations of the effects of conversation on grammar. The puzzle is to find out how it is that complex grammar can emerge from the repeated use of language in conversation.

A crucial step toward understanding how grammar might emerge from the pressures of conversation is to recognize that language operates on multiple timescales at the same time.[26] If we want to see how language is shaped by its environment, we must look at how that shaping can take place at different speeds, simultaneously. Linguists have long known about the *diachronic* or historical timescale, in which languages change gradually over periods of decades and centuries. The environments for historical change in language are human populations and the dynamic processes by which innovations and changes in language can spread.

Another scale for language is the *microgenetic* timescale, the timescale of human thinking. At the microgenetic timescale, the properties of people's brains and minds can play an important role in shaping language, for example through principles of economy in cognitive processing and memory.

And in the *ontogenetic* timescale—the scale of the human lifespan—the ways in which we learn a language, especially during the first few years of life, can affect the way we process and organize language in our minds. A number of researchers have begun to explore how these three timescales are linked and how those links can help us see the full underlying causes of language.[27] This work suggests that if researchers overlook any one of these timescales, we will fail to fully understand why language is the way it is. Each scale exerts its own environmental pressures on the shaping of language.

But there is another timescale that cannot be left out of a full explanation of why language is the way it is. This is the scale that we have focused on in this book. It is called the *enchronic* timescale, the scale at which human conversation operates, the move-by-move exchange of turns at talk, within a cooperative frame of interlocking rights and duties. In the enchronic timescale, the formative environment for language is the high-pressure realm of turn-taking, with its sensitive one-second window for transition. In this environment, people must formulate their turn where there are only milliseconds before the floor may be taken up by another turn hot on its heels. Depending on whether you want to keep hold of the floor or to get the other person to respond, and depending on whether you want to show more or less affiliation with the other person, you will need to design your turn with this in mind.

The connection to language structure is a direct one. To design a turn in conversation, the raw materials we reach for are the grammatical structures and sound structures of the specific language we speak. While all people share the same basic in-built capacities for conversation, to actually *run* the conversation machine we need the moving parts that specific languages provide. Seen from this perspective, the grammatical differences between languages—anything from whether the verb comes first or last in a clause to whether the grammar employs a case-marking system or a verb-agreement system—may have real implications for how turn-taking, repair, and other workings of conversation are organized and carried out.

The direct relation between grammar and conversation is uncharted territory for mainstream linguistics,[28] although the issue is being explored in a specialized subfield called *interactional linguistics.*[29] One of the important shifts that interactional linguists are making in the study of grammar is to move the focus from sentences—which have their home in written language—to other kinds of units, including the smaller fragments that turns at conversation are built from. As we saw in Chapter 3, people listening to conversation are able to tell in advance which bit of speech is likely to be the final part of a turn, allowing them to accurately time their own incoming speech in conversation. To do this, people draw on a range of available information simultaneously, including units of grammatical structure such as phrases, as well as prosodic cues such as stress or lengthening of certain syllables. While much research is still needed to properly describe and understand these units, linguist Stephen Levinson notes that the correct description "must allow for the projectability or predictability of each unit's end—for it is this alone that can account for the recurrent marvels of split-second speaker transition."[30]

Another important shift away from the sentence as the standard unit of language structure is to give more attention to the structure of larger sequences such as the narratives we looked at in Chapter 5. The higher-level directions of conversation not only reveal less-charted dimensions of language structure; they also point to the social and interpersonal commitments that apply when we talk. When researchers shift focus from traditional written forms of language, we begin to see grammar not only as a tool for representing meaning, or for merely structuring information, but also as a tool for organizing social interaction and social life.

I suggested at the start of this chapter that a paradigm shift in the science of language will occur when two competing ideas are brought together: one, that language depends on a uniquely human conversation machine; and two, that language is a private computational system for operating upon information. Of course, if only one of these ideas is correct, then no bridge can be built. This is in fact Chomsky's view. Chomsky and colleagues say that the human capacity for language has no relation to social interaction and that language is only incidentally useful for social interaction: In their words, "language may be used for communication, but that particular function is largely irrelevant."[31] This radical view is untenable, as it denies the well-documented existence of the conversation machine and its inner workings.

A more moderate view, widespread among linguists, is that the two ideas just mentioned are both valid areas of research in their own right, but they happen to have nothing to do with each other. As a result, the formal structures of words and grammatical phrases can be safely studied in isolation from the social functions of turns in conversation, as has largely been the practice to date.

But the truth is more interesting than that. If we acknowledge that language is both a computational system *and* a tool for communication and social action, we can then ask about the relation between these two aspects of language. This opens up new frontiers for research on the human capacity for language.

The research I have described in this book suggests that languages differ in many ways but that they are everywhere supported by the same universal infrastructure for social interaction, a human conversation machine. The workings of this machine are grounded in fundamental properties of language structure, social cognition, and the contexts of interaction.

Three elements of the machine deserve special mention. First, humans are able to commit to social projects and hold each other accountable to these commitments. Without this interpersonal accountability, conversation would not happen. Second, human language lets us use our words to talk about our words. Without this reflexive capacity of language, we would be unable to draw attention to what has been said, and we would be unable to point to transgressions of the rules of conversation or to problems that need repairing. Third, in the flow of interaction, humans interpret every move as being specifically relevant to the one before it and to the one that comes next. This relevance principle provides a glue that binds moves in conversation. In addition, the principle means that we need to keep our contributions in linear order so that we know what is a response to what.

Taken together, these three core elements of the conversation machine provide the conditions under which more specific features of the machine can emerge. For example, both the one-at-a-time rule and the one-second rule in turn-taking emerge from the principles of relevance and interpersonal commitment. The common principles of other-initiated repair in conversation are possible because of the reflexive property of language combined again with the interpersonal commitment that conversation entails.

Scientists of language have much work to do if we are going to truly understand the human capacity for language. Bread-and-butter field research must continue on the complexities of grammar across languages, gathering data on the structure of the world's languages, and continuing work on testing the extent of, and limits on, diversity in human language. Even more urgent is the same kind of documentary research on the workings of *conversation* across very different languages worldwide. What are all of these workings, what are they like, and to what extent do they vary across languages and cultures?

What we have learned about the inner workings of conversation could not have been learned without evidence from the everyday conversations of people around the world. The surprising facts of the universal word "Huh?" would not have been discovered had we not started—like biologists—by going into the field to find out what the beasts we study are really like. It sometimes seems that the things we see in the wilds of conversation inhabit the edges of language, far from the "core," where we find nouns and verbs, subjects and predicates, semantics and syntax. But the truth may be the other way around. It may be that a simple "Huh?" is as close as you can get to the core of the human faculty for language.

ACKNOWLEDGMENTS

What I know about the inner workings of conversation I know from collaborating with many colleagues in projects on social interaction between 2000 and 2014 in the Language and Cognition Group at the Max Planck Institute for Psycholinguistics, Nijmegen, and since then at the University of Sydney. I am especially grateful to members and guests of the Max Planck Institute projects titled Multimodal Interaction, Interactional Foundations of Language, and Human Sociality and Systems of Language Use, and especially Steve Levinson, Paul Kockelman, Bill Hanks, Manny Schegloff, Herb Clark, Betty Couper-Kuhlen, Paul Drew, John Heritage, Mark Dingemanse, Ruth Parry, Seán Roberts, J. P. de Ruiter, Jack Sidnell, and Tanya Stivers. I am grateful to all my co-authors and collaborators whose input is reflected in the studies cited and discussed in this book: Julija Baranova, Joe Blythe, Penny Brown, Mark Dingemanse,

Tyko Dirksmeyer, Paul Drew, Christina Englert, Simeon Floyd, Sonja Gipper, Rosa Gisladottir, Makoto Hayashi, Trine Heinemann, Gertie Hoymann, Kobin Kendrick, Steve Levinson, Lilla Magyari, Elizabeth Manrique, Seán Roberts, Federico Rossano, Giovanni Rossi, Lila San Roque, Francisco Torreira, and Kyung-Eun Yoon.

For help with preparing the figures I am very grateful to Asif Ghazanfar, Felicia Roberts, Seán Roberts, J. P. de Ruiter, Tanya Stivers, and Francisco Torreira. I thank Gus Wheeler and Georgia Carr for their expert assistance in preparing the manuscript for submission.

I would like to thank those who generously read part or all of the manuscript and gave helpful comments: Felicia Roberts, Margaret Enfield, Dan Everett, Hugo Mercier, Ruth Parry, J. P. de Ruiter, Tanya Stivers, Anna Vatanen, and Samantha Williams. At Basic Books, I would like to thank Tisse Takagi and Hélène Barthelemy, and especially T. J. Kelleher for his incisive editing and advice.

I am particularly grateful to my agents, Katinka Matson and Max Brockman. This book would not exist without Katinka's generosity, support, and memorable words of advice. I would also like to thank Paul Bloom for his advice at the start of this project: I wish I had heeded it earlier than I did.

And thank you, as ever, to my family, Na, Nyssa, and Nonnika, for constant love and support.

I dedicate this book to my colleague and mentor Steve Levinson. Without his enduring inspiration and generosity, much of the work discussed in this book would never have happened.

NOTES

Chapter 1: Introduction: What Is Language Like?

1. Darwin 1890: 2.

2. Ibid.: 313.

3. This is fortunately beginning to change, as we will see in later chapters, where we discuss the work of conversation analysts and interactional linguists. For reference materials on this work, see Schegloff 2007, Sidnell 2010, Sidnell and Stivers 2012, and Clift 2016.

4. English is a special exception; it is by far the most closely and extensively studied language in the world, and most good English dictionaries do in fact have entries for "Huh?" and similar words.

5. Darwin 1890: 26.

6. Ibid.: 27.

7. The term *repair* is used in this book with a specific technical meaning, relating to how people in conversation draw attention to problems of speaking, hearing, or understanding and then resolve or "repair" these problems as they go. We

will have much to say about conversational repair later in the book.

8. Chomsky 1965: 3; see my critique of this point in Enfield 2015a.

9. The sociologist Harvey Sacks—a pioneer of the ideas discussed in this book—wrote of a "machinery" for conversation (see McHoul 2005). A comparable term is *interaction engine.* See Enfield and Levinson (2006) and Levinson (2006), in which the idea of an interaction engine is specifically limited to an individual's cognitive capacities, as distinct from an "interaction matrix," the situational framework that introduces universal constraints on the structure of how language is used. With the term *conversation machine* I want to capture the combination of the individual's capacities and the basic situational constraints of human communication. Together, these capacities and constraints mesh to create the machine that runs when we talk.

10. The reader who wants to follow up on this in academic literature may consult the following works and the many references in them: Clark 1996, Sidnell 2010, Sidnell and Stivers 2012, Enfield 2013, Clift 2016.

11. Pinker 1994: 232.

12. See Norman (1988) for an accessible introduction to how this works. See also Clark (1997).

Chapter 2: Conversation Has Rules

1. Searle 1990. See also http://plato.stanford.edu/entries/shared-agency.

2. Gilbert 1992.

3. True cooperation is joint action in this sense. See Michael, Sebanz, and Knoblich 2016.

4. Gilbert 1992: 3.

5. Beach 1996: 120.
6. Atkinson and Drew 1979: 52.
7. Schegloff 1992: 1310.
8. Stivers and Rossano 2010: 6.
9. Rogers and Norton 2011: 139.
10. Clayman and Heritage 2002: 282.
11. Schegloff 1980: 107.
12. Ibid.: 108.
13. Schegloff 2007: 45.
14. Ibid.: 47.
15. Schegloff 1980: 110.
16. Bolden 2006: 665.
17. Sacks 1992 (vol. II): 222–228.
18. www.encyclopaedia.com/pdfs/6/98.pdf.
19. Perry 2003: 113.
20. Garfinkel 1967: 42.
21. Ibid.
22. Ibid.: 43.
23. See Goffman 1963.

Chapter 3: Split-Second Timing

1. See Sidnell 2010, Sidnell and Stivers 2012, Clift 2016.
2. de Ruiter, Mitterer, and Enfield 2006: 516.
3. Levinson and Torreira 2015: 16.
4. Riest, Jorschick, and de Ruiter 2015: 65.
5. I emphasize that the slight gaps and overlaps that we observe do not constitute breaking the turn-taking rules; most gaps or overlaps are extremely brief, and fleeting.
6. http://bionumbers.hms.harvard.edu/bionumber.aspx?s=y&id=100706&ver=0.
7. Levelt 1989.
8. Indefrey and Levelt 2004.

9. Levinson 2016: 8.

10. Duncan 1974, Duncan and Niederehe 1974. The Sacks and colleagues 1974 paper was partly framed as a response to Duncan's work.

11. Beattie, Cutler, and Pearson 1982.

12. de Ruiter, Mitterer, and Enfield 2006.

13. See Bögels and Torreira 2015: 55.

14. Bögels and Torreira 2015.

15. See Ford and Thompson 1996, who also argue that there are multiple cues for turn transition.

16. This suggests that 68 percent of people tested are poor projectors because they missed the cue from the spliced short sentence extract. The authors of the study suggest that this is possibly an effect of the fact that the sentences in the experiment are so short. The lack of prior context for these sentences makes it difficult for listeners in the experiment to gear up to project with their usual accuracy. What's important here is that so many people did project a turn ending in the middle of this question, where none did in the nonspliced condition.

17. Lehtonen and Sajavaara 1985: 198.

18. Reisman 1974.

19. Ibid.

20. Tannen 1984.

21. See Stivers et al. 2009, Enfield, Stivers, and Levinson 2010. The research team members also included Penelope Brown, Christina Englert, Makoto Hayashi, Trine Heinemann, Gertie Hoymann, Federico Rossano, Jan Peter de Ruiter, Kyung-Eun Yoon, and Stephen Levinson.

22. Takahashi, Narayanan, and Ghazanfar 2013.

23. Ibid.: 2162.

24. Ibid.: 2165.

Chapter 4: The One-Second Window

1. Pomerantz 1984: 77.

2. Levinson 1983: 335.

3. See Pomerantz and Heritage 2012; on questions, see Raymond 2003.

4. Jefferson 1989. Herbert Clark later noted the same phenomenon (1996: 268).

5. See Egeth 1966, Bamber 1969, Bindra, Donderi, and Nishisato 1968, Ratcliff 1987.

6. Egeth 1966: 249–250.

7. 72% = 130/183; Stivers 2010: 2779.

8. Clark and Fox Tree 2002: 84.

9. Roberts and Francis 2013.

10. Ibid.: 476.

11. What I am calling the "on-time zone" for change to a new speaker—the period of around a half-second from the end of a person's turn—is similar to what is referred to by conversation analysts as the "transition space" (see Clayman 2013).

12. Atkinson and Drew 1979: 58.

13. A "particle" is a word like "well," "um," or "so"; these words are dedicated to managing the interaction rather than to providing information about the topic that is being spoken about. Blakemore (1987) describes particles as having "procedural" meaning (as opposed to the "conceptual" meaning that words like nouns and verbs have).

14. Kendrick and Torreira 2015: 273.

15. Ibid.: 22.

16. Ibid.: 23.

17. Stivers et al. 2009: 10589.

18. Roberts, Margutti, and Takano 2011.

19. For details, see the figure on page 343 of Roberts, Margutti, and Takano 2011.

Chapter 5: Traffic Signals

1. Clark and Fox Tree 2002: 73.
2. Ibid.: 74.
3. Ibid.
4. Levelt 1989: 480–481.
5. Goffman 1981: 293.
6. Schegloff 2010: 142.
7. Ibid.
8. Ibid.: 142–143.
9. Ibid.: 143–144.
10. Ibid.: 147–148.
11. Ibid.: 149.
12. Ibid.: 149–150.
13. Ibid.: 151.
14. Ibid.: 158.
15. Ibid.: 142. Giving a reason for a call could conceivably be regarded as a dispreferred move because it entails shutting down one line of conversation in order to move to the business phase.
16. http://languagelog.ldc.upenn.edu/nll/?p=13713.
17. Jefferson 1974: 184.
18. Levelt 1989: 484.
19. Schegloff 1982: 84.
20. Clark 1996.
21. Schegloff 1982.
22. Clark 1994: 1006–1007.
23. Bavelas, Coates, and Johnson 2000.
24. Linguist Charles Goodwin (1986) distinguishes between continuers and assessments. Continuers, including

"mm-hmm" and "uh-huh," are used simply to signal that the other should continue. By contrast, assessments, including "Wow" and "Oh my God," convey an appraisal of what has been said, and rather than urging continuation, they tend to occur when someone is finishing a narrative.

25. Examples from Bavelas, Coates, and Johnson 2000: 943.

26. Ibid.: 943–944.

27. Bavelas, Coates, and Johnson 2000: 948.

28. Rovee and Rovee 1969.

29. Murray and Trevarthen 1986: 15–29. See also Striano et al. 2006.

30. Murray and Trevarthen 1986: 24–25.

31. Tyack 2003: 360.

32. See Giles 1991.

33. Turner and West 2010.

34. Tomasello 2016.

35. On "so," see Bolden 2006; on "oh," see Heritage 2002; on "okay," see Bangerter and Clark 2003; for these "discourse markers" more generally, see Schiffrin 1988, Wierzbicka 2003.

Chapter 6: The Glue of Relevance

1. Haimoff 1981: 144.

2. Rossano 2013: 167.

3. The word *relevance* has a technical meaning in linguistic research. There is a major field of study known as *relevance theory* (Sperber and Wilson 1995). See also Grice 1989 and Levinson 2000.

4. Sperber and Wilson 1995: 34.

5. See Garfinkel 1967.

6. Zeitlyn 1995: 199.

7. Levinson 1995: 236.

8. Clark 1979: 430.

9. Ibid.: 460.

10. Ibid.: 463.

11. Ibid.: 464.

12. Melis et al. 2016: 1.

13. "Supporting the cultural intelligence hypothesis and contradicting the hypothesis that humans simply have more 'general intelligence,' we found that the children and chimpanzees had very similar cognitive skills for dealing with the physical world but that the children had more sophisticated cognitive skills than either of the ape species for dealing with the social world" (Herrmann et al. 2007: 1360).

Chapter 7: Repair

1. Schegloff, Jefferson, and Sacks 1977: 367.

2. Jefferson 1978a, 1978b, Heritage 1984.

3. Dingemanse et al. 2015, Dingemanse and Enfield 2015.

4. Conversation analysts use the term *progressivity* to describe the forward-moving quality of conversation (see Stivers and Robinson 2006).

5. Schegloff, Jefferson, and Sacks 1977.

6. Ibid.: 369.

7. Ibid.: 378.

8. Clark and Schaefer 1987.

9. Ibid.

10. Ibid.: 23.

11. See Schegloff 2007 for a technical introduction to the organization of sequences in English conversation.

12. See Dingemanse et al. 2015, Figure 2, page 5.

13. See the 2015 issue of the journal *Open Linguistics* for parallel examples from Cha'palaa, Dutch, English, Icelandic, Italian, Lao, Sign Language of Argentina, Mandarin Chinese, Murrinhpatha, Russian, Siwu, and Yélî Dnye.

14. Both "hmm" and "Huh" can have other meanings, especially when they are pronounced with a different intonation. When pronounced with a falling tone, "hmm!" can mean "I'm skeptical," while "Huh!" can mean "That's interesting." I do not discuss those other meanings here.

15. See Bloomfield 1933, Kockelman 2003.

16. Example from Floyd 2015.

17. Example from Dingemanse 2015.

18. Example from Rossi 2015.

19. See Svennevig 2008 on Norwegian, Kim 1999 on Korean.

20. Drew 1997: 76.

21. Data from Elizabeth Holt: ref. Holt:X:C:1:1:1:287 (06.20).

22. See Dingemanse et al. 2015, Figure 5, page 9.

Chapter 8: The Universal Word: "Huh?"

1. Lao example from Enfield 2015b; Siwu example from Dingemanse 2015; Cha'palaa example provided from Simeon Floyd's data (CHSF2011_02_15S5_1667143). See Floyd 2015.

2. See Wierzbicka 1996, Dixon 2004, Hauser, Chomsky, and Fitch 2002.

3. With every claim of universality in linguistics, there is a claim to have found counterevidence. See, for example, Everett 2005, 2009, 2012, Nevins, Pesetsky, and Rodrigues 2009a, 2009b.

4. See Enfield et al. 2013.

5. The Spanish data were supplied by Francisco Torreira; the Cha'palaa data were supplied by Simeon Floyd.

6. See, for example, Mazeland (1987: 3), Schegloff (1997: 506), Ward (2006: 129).

7. Dingemanse, Torreira, and Enfield 2013.

8. See huh.ideophone.org.

9. Tylor 1889: 272.

10. We can never rule out the possibility of a counterexample coming to light one day. This is a hallmark of the inductive empirical science of linguistics. It encourages you to keep going back out into the field.

11. Saussure 1916.

12. Liu et al. 2010.

13. Kendrick 2015: 151.

14. See huh.ideophone.org for a collection of press stories on the issue.

Chapter 9: Conclusion: The Capacity for Language

1. There are many reasons to think this: Dąbrowska 2004, Levinson and Evans 2010, Everett 2012, Dor 2015, Evans 2015, Christiansen and Chater 2016.

2. See Levinson 2006, Dor, Knight, and Lewis 2014.

3. For recent discussions of the concept of Universal Grammar, see Hauser, Chomsky, and Fitch 2002, Evans and Levinson 2009, Everett 2012. Chomsky now says that Universal Grammar consists of the combination of the words of language plus a single powerful syntactic rule called "merge" (Chomsky 2012: 41). The term *Universal Grammar* is best used exclusively with the "narrow sense" meaning, pointing to a language-specific mental device. Confusingly, it is also sometimes used in the broad sense (Chomsky 2011: 264).

4. Neil Smith makes these points from 8m55s here: www.theguardian.com/science/audio/2017/jan/11/universal-grammar-are-we-born-knowing-the-rules-of-language-science-weekly-podcast.

5. Hauser, Chomsky, and Fitch 2002, Christiansen and Chater 2016, Everett 2005.

6. While many such principles of Universal Grammar have been proposed over the years, Chomsky has now abandoned

all but the principle of recursion, as embodied in a grammatical rule called "merge," which allows people to take two structures and combine them into a single one. See Hauser, Chomsky, and Fitch 2002, Chomsky 2012: 41.

7. Goldberg 2006, Bybee 2010, Everett 2012, Langacker 1987, Evans and Levinson 2009, Prinz 2012.

8. Ansaldo and Enfield 2016, Goldberg 2006, Bybee 2010, Dąbrowska 2004. See commentaries on Evans and Levinson 2009, and see Everett 2012, Langacker 1987, Croft 2001, Evans 2015, Christiansen and Chater 2016.

9. Prinz 2012, Everett 2012, Evans 2015.

10. Langacker (1987) is an influential statement of the view that language is supported in our species by knowledge that is not specific to language. See the long list of spirited commentaries in Evans and Levinson 2009; see also Everett 2012, Evans 2015, and the many references in those sources.

11. This does not detract from the enormous value of the work on both sides.

12. See functionalists from Chafe 1994 to Lambrecht 1994 to Halliday 1994 to Martin and Rose 2007.

13. Any professional linguist can of course say that "Huh?" is classified as an interjection. However, this traditionally defines it as a word that does not really have any grammar at all (see Bloomfield 1933, Kockelman 2003). As such, "Huh?" is relegated to the margins of language.

14. Chomsky 1965: 3; see also Enfield 2015a and Clark and Fox Tree 2002.

15. In his 2004 book *Doctor Dolittle's Delusion*, the linguist Stephen Anderson explores what makes language possible in our species. The exclusive focus is syntax—the principles for organizing phrases and sentences—and there are no references in the index to conversation, repair, turn-taking, cooperation,

or relevance. In linguist Ray Jackendoff's 2002 tour de force *Foundations of Language,* just 3 pages out of nearly 500 are dedicated to "discourse," that is, structure in language beyond the level of the sentence. Even the influential 2009 article "The Myth of Language Universals" by Nick Evans and Steve Levinson, with its close attention to language diversity and the contexts of language, has little mention of conversation, repair, turn-taking, cooperation, or relevance.

16. Here I am not talking about the volumes of research on language in social context in fields such as sociology (Sidnell and Stivers 2012) and anthropology (Enfield, Kockelman, and Sidnell 2014). I am talking about research on the question of our species' capacity for language.

17. Dunbar 1993, 1996.

18. Hurford 1999: 182.

19. See summary diagram on page 294 of Tomasello 2008.

20. Christiansen and Chater 2016.

21. Christiansen and Chater 2008.

22. My published commentary on Chater and Christiansen's article is titled "Language as Shaped by Social Interaction." There I suggest that "co-contingency of unit contribution and response may be argued to serve as a direct determinant shaping linguistic organization" (Enfield 2008: 520). See also Roberts and Levinson (2017), who remark that "the greatest functional pressures on language structure are likely to come from the very special circumstances in which it is primarily used. That special niche is conversation." See also Roberts and Levinson 2015. For pioneering work, see Schegloff 1989.

23. See Thompson, Fox, and Couper-Kuhlen 2015, Schegloff, Ochs, and Thompson 1996.

24. Roberts and Levinson 2015, 2017. See pioneering work: Thompson 1998 and Schegloff 1989.

25. This work is beginning (see Dingemanse and Enfield 2015: 97–98 for references). Interactional linguists have published research on numerous languages including German, Swedish, Finnish, and Japanese—that said, the current scope of diversity is a tiny fraction of what informs the study of grammar across human languages.

26. For a technical sketch of this idea, see my book *Natural Causes of Language* (Enfield 2014). See also Christiansen and Chater 2016.

27. Enfield 2014, Steffenson and Fill 2013, Uryu, Steffenson, and Kramsch 2014, Rączaszek-Leonardi 2010, Christiansen and Chater 2016.

28. These horizons have recently been prospected by Stephen Levinson and members of his pioneering INTERACT project, held at the Max Planck Institute for Psycholinguistics, 2011–2015. See Roberts and Levinson 2015, 2017.

29. See, for example, Thompson, Fox, and Couper-Kuhlen 2015.

30. Levinson 1983: 140. Stephen Levinson has recently overseen groundbreaking research on the projection problem (see Holler et al. 2015 on turn-taking).

31. Bolhuis et al. 2014. See also Piantadosi, Tily, and Gibson 2012.

BIBLIOGRAPHY

Anderson, Stephen R. 2004. *Doctor Dolittle's Delusion*. New Haven: Yale University Press.

Ansaldo, Umberto, and N. J. Enfield, eds. 2016. *Is the Language Faculty Nonlinguistic?* Lausanne, Switzerland: Frontiers Media.

Atkinson, J. Maxwell, and Paul Drew. 1979. *Order in Court*. Atlantic Highlands, NJ: Humanities Press.

Bamber, Donald. 1969. "Reaction Times and Error Rates for 'Same'-'Different' Judgements of Multidimensional Stimuli." *Perception and Psycholinguistics* 6 (3): 167–174.

Bangerter, Adrian, and Herbert H. Clark. 2003. "Navigating Joint Projects with Dialogue." *Cognitive Science* 27: 195–225.

Bavelas, Janet, Linda Coates, and Trudy Johnson. 2000. "Listeners as Co-Narrators." *Journal of Personality and Social Psychology* 78 (6): 941–952.

Beach, Wayne A. (1996). *Conversations About Illness: Family Preoccupations with Bulimia*. Mahwah, NJ: Erlbaum.

Beattie, Geoffrey, Anne Cutler, and Mark Pearson. 1982. "Why Is Mrs. Thatcher Interrupted So Often?" *Nature* 300: 744–747.

Bindra, Dalbir, Don C. Donderi, and Shizuhiko Nishisato. 1968. "Decision Latencies of 'Same' and 'Different' Judgements." *Perception and Psychophysics* 3: 121–136.

Blakemore, Diane. 1987. *Semantic Constraints on Relevance*. Oxford: Blackwell.

Bloomfield, Leonard. 1933. *Language*. New York: Holt.

Bögels, Sara, and Francisco Torreira. 2015. "Listeners Use Intonational Phrase Boundaries to Project Turn Ends in Spoken Interaction." *Journal of Phonetics* 52: 46–57.

Bolden, Galina B. 2006. "Little Words That Matter: Discourse Markers 'So' and 'Oh' and the Doing of Other-Attentiveness in Social Interaction." *Journal of Communication* 56: 661–688.

Bolhuis, Johan J., Ian Tattersall, Noam Chomsky, and Robert C. Berwick. 2014. "How Could Language Have Evolved?" *PLoS Biology* 12 (8). doi:10.1371/journal.pbio.1001934.

Bybee, Joan. 2010. *Language, Usage and Cognition*. Cambridge: Cambridge University Press.

Chafe, Wallace. 1994. *Discourse, Consciousness, and Time: The Flow and Displacement of Conscious Experience in Speaking and Writing*. Chicago: University of Chicago Press.

Chomsky, Noam A. 1965. *Aspects of the Theory of Syntax*. Cambridge: MIT Press.

Chomsky, Noam A. 2011. "Language and Other Cognitive Systems: What Is Special About Language?" *Language Learning and Development* 7 (4): 263–278. doi:10.1080/15475441.2011.584041.

Chomsky, Noam A. 2012. *The Science of Language*. Cambridge: Cambridge University Press.

Christiansen, Morten H., and Nick Chater. 2008. "Language as Shaped by the Brain." *Behavioral and Brain Sciences* 31 (5): 489–509.

Christiansen, Morten H., and Nick Chater. 2016. *Creating Language: Integrating Evolution, Acquisition, and Processing.* Cambridge: MIT Press.

Clark, Andy. 1997. *Being There: Putting Brain, Body, and World Together Again.* Cambridge: MIT Press.

Clark, Herbert H. 1979. "Responding to Indirect Speech Acts." *Cognitive Psychology* 11: 430–477.

Clark, Herbert H. 1994. "Discourse in Production." In *Handbook of Psycholinguistics,* edited by M. A. Gernsbacher, 985–1021. San Diego: Academic Press.

Clark, Herbert H. 1996. *Using Language.* Cambridge: Cambridge University Press.

Clark, Herbert H., and Jean E. Fox Tree. 2002. "Using Uh and Um in Spontaneous Speaking." *Cognition* 84: 73–111.

Clark, Herbert H., and E. F. Schaefer. 1987. "Contributing to Discourse." *Cognitive Science* 13: 259–294.

Clayman, Steven. 2013. "Turn-Constructional Units and the Transition-Relevance Place." In *The Handbook of Conversation Analysis,* edited by Jack Sidnell and Tanya Stivers, 150–166. Hoboken, NJ: Blackwell.

Clayman, Steven, and John Heritage. 2002. *The News Interview: Journalists and Public Figures on the Air.* New York: Cambridge University Press.

Clift, Rebecca. 2016. *Conversation Analysis.* New York: Cambridge University Press.

Croft, William. 2001. *Radical Construction Grammar: Syntactic Theory in Typological Perspective.* Oxford: Oxford University Press.

Dąbrowska, Ewa. 2004. *Language, Mind and Brain: Some Psychological and Neurological Constraints on Theories of Grammar.* Edinburgh: Edinburgh University Press.

Darwin, Charles. 1890. *The Formation of Vegetable Mould Through the Action of Worms, with Observations on Their Habits*. New York: Appleton.

de Ruiter, Jan Peter, Holger Mitterer, and N. J. Enfield. 2006. "Projecting the End of a Speaker's Turn: A Cognitive Cornerstone of Conversation." *Language* 82 (3): 515–535.

Dingemanse, Mark. 2015. "Other-Initiated Repair in Siwu." *Open Linguistics* 1: 232–255. doi:10.1515/opli-2015-0001.

Dingemanse, Mark, and N. J. Enfield. 2015. "Other-Initiated Repair Across Languages: Towards a Typology of Conversational Structures." *Open Linguistics* 1: 96–118. doi:10.2478/opli-2014-0007.

Dingemanse, Mark, Sean G. Roberts, Julija Baranova, Joe Blythe, Paul Drew, Simeon Floyd, Rosa S. Gisladottir, et al. 2015. "Universal Principles in the Repair of Communication Problems." *PLoS ONE* 10 (9): e0136100. doi:10.1371/journal.pone.0136100.

Dingemanse, Mark, Francisco Torreira, and N. J. Enfield. 2013. "Is 'Huh?' a Universal Word? Conversational Infrastructure and the Convergent Evolution of Linguistic Items." *PLoS ONE* 8 (11): e78273. doi:10.1371/journal.pone.0078273.

Dixon, R. M. W. 2004. "Adjective Classes in Typological Perspective." In *Adjective Classes: A Cross-Linguistic Typology,* edited by R. M. W. Dixon and Alexandra Y. Aikhenvald, 1–49. Oxford: Oxford University Press.

Dor, Daniel. 2015. *The Instruction of Imagination: Language as a Social Communication Technology*. Oxford: Oxford University Press.

Dor, Daniel, Chris Knight, and J. Lewis, eds. 2014. *The Social Origins of Language: Studies in the Evolution of Language*. Oxford: Oxford University Press.

Drew, Paul. 1997. "'Open' Class Repair Initiators in Response to Sequential Sources of Trouble in Conversation." *Journal of Pragmatics* 28: 69–101.

Dunbar, Robin I. M. 1993. "Coevolution of Neocortical Size, Group Size, and Language in Humans." *Behavioral and Brain Sciences* 16: 681–735.

Dunbar, Robin I. M. 1996. *Grooming, Gossip and the Evolution of Language*. London: Faber and Faber.

Duncan, Starkey. 1974. "On the Structure of Speaker-Auditor Interaction During Speaking Turns." *Language in Society* 3 (2): 161–180.

Duncan, Starkey, and G. Niederehe. 1974. "On Signalling That It's Your Turn to Speak." *Journal of Experimental Social Psychology* 10 (3): 234–247.

Egeth, Howard. 1966. "Parallel Versus Serial Processes in Multidimensional Stimulus Discrimination." *Perception and Psychophysics* 1: 245–252.

Enfield, N. J. 2008. "Language as Shaped by Social Interaction [Commentary on Christiansen and Chater]." *Behavioral and Brain Sciences* 31 (5): 519–520. doi:10.1017/S0140525X08005104.

Enfield, N. J. 2013. *Relationship Thinking: Agency, Enchrony, and Human Sociality*. New York: Oxford University Press.

Enfield, N. J. 2014. *Natural Causes of Language: Frames, Biases, and Cultural Transmission*. Berlin: Language Science Press.

Enfield, N. J. 2015a. "A Science of Language Should Deal Only with 'Competence.'" In *This Idea Must Die: Scientific Theories That Are Blocking Progress,* edited by John Brockman, 243–244. New York: Harper Perennial.

Enfield, N. J. 2015b. "Other-Initiated Repair in Lao." *Open Linguistics* 1: 119–144.

Enfield, N. J., Mark Dingemanse, Julija Baranova, Joe Blythe, Penelope Brown, Tyko Dirksmeyer, Paul Drew, et al. 2013. "Huh? What?—A First Survey in 21 Languages." In *Conversational Repair and Human Understanding*, edited by Makoto Hayashi, Geoffrey Raymond, and Jack Sidnell, 30: 343–380. Studies in Interactional Sociolinguistics. New York: Cambridge University Press.

Enfield, N. J., Paul Kockelman, and Jack Sidnell, eds. 2014. *The Cambridge Handbook of Linguistic Anthropology*. Cambridge: Cambridge University Press.

Enfield, N. J., and Stephen C. Levinson. 2006. "Introduction: Human Sociality as a New Interdisciplinary Field." In *Roots of Human Sociality: Culture, Cognition, and Interaction*, edited by N. J. Enfield and Stephen C. Levinson, 1–38. Oxford: Berg.

Enfield, N. J., Tanya Stivers, and Stephen C. Levinson, eds. 2010. "Question-Response Sequences in Conversation Across Ten Languages." Special issue of *Journal of Pragmatics* 42 (10).

Evans, Nicholas D., and Stephen C. Levinson. 2009. "The Myth of Language Universals: Language Diversity and Its Importance for Cognitive Science." *Behavioral and Brain Sciences* 32 (5): 429–448.

Evans, Vyvyan. 2015. *The Crucible of Language: How Language and Mind Create Meaning*. Cambridge: Cambridge University Press.

Everett, Daniel L. 2005. "Cultural Constraints on Grammar and Cognition in Pirahã." *Current Anthropology* 46 (4): 621–646.

Everett, Daniel L. 2009. "Pirahã Culture and Grammar: A Response to Some Criticisms." *Language* 85 (2): 405–442.

Everett, Daniel L. 2012. *Language: The Cultural Tool*. London: Profile.

Floyd, Simeon. 2015. "Other-Initiated Repair in Cha'palaa." *Open Linguistics* 1 (1): 467–489. doi:10.1515/opli-2015-0014.

Ford, C. E., and S. A. Thompson. 1996. "Interactional Units in Conversation: Syntactic, Intonational, and Pragmatic Resources for the Management of Turns." *Studies in Interactional Sociolinguistics* 13: 134–184.

Garfinkel, Harold. 1967. *Studies in Ethnomethodology*. Englewood Cliffs, NJ: Prentice-Hall.

Gilbert, Margaret. 1992. *On Social Facts*. Princeton, NJ: Princeton University Press.

Giles, Howard. 1991. "Accommodation Theory: Communication Context and Consequence." In *Contexts of Accommodation*, edited by Howard Giles, Justine Coupland, and N. Coupland, 1–68. New York: Cambridge University Press.

Goffman, Erving. 1963. *Stigma: Notes on the Management of Spoiled Identity*. New York: Touchstone.

Goffman, Erving. 1981. *Forms of Talk*. Philadelphia: University of Pennsylvania Press.

Goldberg, Adele E. 2006. *Constructions at Work: The Nature of Generalization in Language*. Oxford: Oxford University Press.

Goodwin, Charles. 1986. "Between and Within: Alternative Treatments of Continuers and Assessments." *Human Studies* 9: 205–217.

Grice, H. Paul. 1989. *Studies in the Way of Words*. Cambridge: Harvard University Press.

Haimoff, Elliott H. 1981. "Video Analysis of Siamang (Hylobates Syndactylus) Songs." *Behaviour* 76 (1/2): 128–151.

Halliday, M. A. K. 1994. *Introduction to Functional Grammar*. 2nd ed. London: Edward Arnold.

Hauser, Marc D., Noam Chomsky, and W. Tecumseh Fitch. 2002. "The Faculty of Language: What Is It, Who Has It, and How Did It Evolve." *Science* 298: 1569–1579.

Heritage, John. 1984. *Garfinkel and Ethnomethodology*. Cambridge: Polity.

Heritage, John. 2002. "Oh-Prefaced Responses to Assessments: A Method of Modifying Agreement/Disagreement." In *The Language of Turn and Sequence*, edited by C. E. Ford, Barbara Fox, and Sandra A. Thompson, 196–224. New York: Oxford University Press.

Herrmann, Esther, Josep Call, María Victoria Hernández-Lloreda, Brian Hare, and Michael Tomasello. 2007. "Humans Have Evolved Specialized Skills of Social Cognition: The Cultural Intelligence Hypothesis." *Science* 317: 1360–1366.

Holler, Judith, Kobin H. Kendrick, Marisa Casillas, and Stephen C. Levinson, eds. 2015. *Turn-Taking in Human Communicative Interaction*. Lausanne, Switzerland: Frontiers Media.

Hurford, James R. 1999. "The Evolution of Language and Languages." In *The Evolution of Culture*, 173–193. Edinburgh: Edinburgh University Press.

Indefrey, Peter, and Willem J. M. Levelt. 2004. "The Spatial and Temporal Signatures of Word Production Components." *Cognition* 92: 101–144. doi:10.1016/j.cognition.2002.06.001.

Jackendoff, Ray. 2002. *Foundations of Language: Brain, Meaning, Grammar, Evolution*. Oxford: Oxford University Press.

Jefferson, Gail. 1974. "Error Correction as an Interactional Resource." *Language in Society* 2: 181–199.

Jefferson, Gail. 1978a. "Sequential Aspects of Storytelling in Conversation." In *Studies in the Organization of Conversational Interaction*, edited by Jim Schenkein, 219–248. New York: Academic Press.

Jefferson, Gail. 1978b. "What's in a 'Nyem'?" *Sociology* 1 (1): 135–139.

Jefferson, Gail. 1989. "Preliminary Notes on a Possible Metric Which Provides for a 'Standard Maximum' Silence of Approximately One Second in Conversation." In *Conversation: An Interdisciplinary Perspective,* edited by D. Roger and P. Bull, 166–196. Clevedon: Multilingual Matters.

Kendrick, Kobin H. 2015. "The Intersection of Turn-Taking and Repair: The Timing of Other-Initiations of Repair in Conversation." *Frontiers in Psychology* 6: 250. doi:doi:10.3389/fpsyg.2015.00250.

Kendrick, Kobin H., and Francisco Torreira. 2015. "The Timing and Construction of Preference: A Quantitative Study." *Discourse Processes* 52 (4): 255–289.

Kim, Kyu-hyunn. 1999. "Other-Initiated Repair Sequences in Korean Conversation: Types and Functions." *Discourse and Cognition* 6: 141–168.

Kockelman, Paul. 2003. "The Meanings of Interjections in Q'eqchi' Maya: From Emotive Reaction to Social and Discursive Action." *Current Anthropology* 44 (4): 467–490.

Lambrecht, Knud. 1994. *Information Structure and Sentence Form: Topic, Focus and the Mental Representations of Discourse Referents/Grammatical Relations.* Cambridge: Cambridge University Press.

Langacker, Ronald W. 1987. *Foundations of Cognitive Grammar: Volume I, Theoretical Prerequisites.* Stanford: Stanford University Press.

Lehtonen, Jaakko, and Kari Sajavaara. 1985. "The Silent Finn." In *Perspectives on Silence,* edited by Deborah Tannen and Muriel Saville-Troike, 193–204. Norwood, NJ: Ablex.

Levelt, Willem J. M. 1989. *Speaking: From Intention to Articulation.* Cambridge: MIT Press.

Levinson, Stephen C. 1983. *Pragmatics.* Cambridge: Cambridge University Press.

Levinson, Stephen C. 1995. "Interactional Biases in Human Thinking." In *Social Intelligence and Interaction: Expressions and Implications of the Social Bias in Human Intelligence*, edited by Esther N. Goody, 221–260. Cambridge: Cambridge University Press.

Levinson, Stephen C. 2000. *Presumptive Meanings: The Theory of Generalized Conversational Implicature*. Cambridge: MIT Press.

Levinson, Stephen C. 2006. "On the Human 'Interaction Engine.'" In *Roots of Human Sociality: Culture, Cognition and Interaction*, edited by N. J. Enfield and Stephen C. Levinson, 39–69. Oxford: Berg.

Levinson, Stephen C. 2016. "Turn-Taking in Human Communication—Origins and Implications for Language Processing." *Trends in Cognitive Sciences* 20 (1): 6–14.

Levinson, S. C., and N. Evans. 2010. "Time for a Sea-Change in Linguistics: Response to Comments on 'The Myth of Language Universals.'" *Lingua* 120: 2733–2758.

Levinson, Stephen C., and Francisco Torreira. 2015. "Timing in Turn-Taking and Its Implications for Processing Models of Language." *Frontiers in Psychology* 6 (731): 10–26.

Liu, Y., J. A. Cotton, B. Shen, X. Han, S. J. Rossiter, and S. Zhang. 2010. "Convergent Sequence Evolution Between Echolocating Bats and Dolphins." *Current Biology* 20 (2): R53–R54. doi:10.1016/j.cub.2009.11.058.

Martin, J. R., and David Rose. 2007. *Working with Discourse: Meaning Beyond the Clause*. London: Continuum.

Mazeland, Harrie. 1987. "A Short Remark on the Analysis of Institutional Interaction: The Organization of Repair in Lessons." In *International Pragmatics Association (IPrA) Conference Proceedings*. Antwerp.

McHoul, Alec. 2005. Aspects of Aspects: On Harvey Sacks's "Missing" Book, *Aspects of the Sequential Organization of Conversation* (1970). Human Studies 28, 113–128.

Melis, Alicia P., Patricia Grocke, Josefine Kalbitz, and Michael Tomasello. 2016. "One for You, One for Me: Humans' Unique Turn-Taking Skills." *Psychological Science OnlineFirst*, 1–10. doi:10.1177/0956797616644070.

Michael, John, Natalie Sebanz, and Günther Knoblich. 2016. "The Sense of Commitment: A Minimal Approach." *Frontiers in Psychology* 6 (1968). doi:10.3389/fpsyg.2015.01968.

Murray, Lynne, and Colwyn Trevarthen. 1986. "The Infant's Role in Mother-Infant Communications." *Journal of Child Language* 13 (1): 15–29. doi:https://doi.org/10.1017/S0305000900000271.

Nevins, Andrew, David Pesetsky, and Cilene Rodrigues. 2009a. "Evidence and Argumentation: A Reply to Everett (2009)." *Language* 85 (3): 671–681.

Nevins, Andrew, David Pesetsky, and Cilene Rodrigues. 2009b. "Pirahã Exceptionality: A Reassessment." *Language* 85: 355–404.

Norman, Donald A. 1988. *The Design of Everyday Things*. New York: Basic Books.

Perry, Susan. 2003. "Coalitionary Aggression in White-Faced Capuchins." In *Animal Social Complexity: Intelligence, Culture, and Individualized Societies*, edited by Frans B. M. de Waal and Peter L. Tyack, 111–114. Cambridge: Harvard University Press.

Piantadosi, Steven T., Harry Tily, and Edward Gibson. 2012. "The Communicative Function of Ambiguity in Language." *Cognition* 122: 28–129.

Pinker, Steven. 1994. *The Language Instinct: How the Mind Creates Language*. New York: William Morrow.

Pomerantz, Anita. 1984. "Agreeing and Disagreeing with Assessments: Some Features of Preferred/Dispreferred Turn Shapes." In *Structures of Social Action: Studies in Conversation Analysis,* edited by J. Maxwell Atkinson and John Heritage, 57–101. Cambridge: Cambridge University Press.

Pomerantz, Anita, and John Heritage. 2012. "Preference." In *The Handbook of Conversation Analysis,* edited by Jack Sidnell and Tanya Stivers, 210–228. Oxford: Wiley-Blackwell.

Prinz, Jesse J. 2012. *Beyond Human Nature: How Culture and Experience Shape Our Lives.* London: Allen Lane.

Rączaszek-Leonardi, J. 2010. "Multiple Time-Scales of Language Dynamics: An Example from Psycholinguistics." *Ecological Psychology* 22 (4): 269–285.

Ratcliff, Roger. 1987. "More on the Speed and Accuracy of Positive and Negative Responses." *Psychological Review* 94 (2): 277–280.

Raymond, Geoffrey. 2003. "Grammar and Social Organization: Yes/No Interrogatives and the Structure of Responding." *American Sociological Review* 68: 939–967.

Reisman, Karl. 1974. "Contrapuntal Conversations in an Antiguan Village." In *Explorations in the Ethnography of Speaking,* edited by Richard Bauman and Joel Sherzer, 110–124. Cambridge: Cambridge University Press.

Riest, Carina, Annette B. Jorschick, and Jan P. de Ruiter. 2015. "Anticipation in Turn-Taking: Mechanisms and Information Sources." *Frontiers in Psychology* 6 (89): 62–75. doi:https://doi.org/10.3389/fpsyg.2015.00089.

Roberts, Felicia, and Alexander L. Francis. 2013. "Identifying a Temporal Threshold of Tolerance for Silent Gaps After Requests." *Journal of the Acoustic Society of America* 133 (6): 471–477.

Roberts, Felicia, Piera Margutti, and Shoji Takano. 2011. "Judgements Concerning the Valence of Inter-Turn Silence Across Speakers of American English, Italian, and Japanese." *Discourse Processes* 48 (5): 331–354.

Roberts, Seán G., and Stephen C. Levinson. 2015. "On-Line Pressures for Turn-Taking Constrain the Cultural Evolution of Word Order." In *Workshop on Cognitive Linguistics and the Evolution of Language*. Newcastle University, UK.

Roberts, Seán G., and Stephen C. Levinson. 2017. "Conversation, Cognition and Cultural Evolution: A Model of the Cultural Evolution of Word Order Through Pressures Imposed from Turn Taking in Conversation," edited by S. Hartmann, M. Pleyer, J. Winters, and J. Zlatev. *Interaction Studies* (special issue on Interaction and Iconicity in the Evolution of Language).

Rogers, T., and M. I. Norton. 2011. "The Artful Dodger: Answering the Wrong Question the Right Way." *Journal of Experimental Psychology: Applied* 17 (2): 139–147.

Rossano, Federico. 2013. "Sequence Organization and Timing of Bonobo Mother-Infant Interactions." *Interaction Studies* 14 (2): 160–189.

Rossi, Giovanni. 2015. "Other-Initiated Repair in Italian." *Open Linguistics* 1: 256–282.

Rovee, C. K., and D. T. Rovee. 1969. "Conjugate Reinforcement of Infant Exploratory Behavior." *Journal of Experimental Child Psychology* 8: 33–39.

Sacks, Harvey. 1992. *Lectures on Conversation*. London: Blackwell.

Sacks, Harvey, Emanuel A. Schegloff, and Gail Jefferson. 1974. "A Simplest Systematics for the Organization of Turn-Taking for Conversation." *Language* 50 (4): 696–735.

Saussure, Ferdinand de. 1916. *Cours De Linguistique Générale.* Paris: Payot.

Schegloff, Emanuel A. 1980. "Preliminaries to Preliminaries: 'Can I Ask You a Question?'" Edited by D. Zimmerman and C. West. *Sociological Inquiry* 50 (3–4): 104–152.

Schegloff, Emanuel A. 1982. "Discourse as an Interactional Achievement: Some Uses of 'Uh Huh' and Other Things That Come Between Sentences." In *Georgetown University Roundtable on Languages and Linguistics 1981; Analyzing Discourse: Text and Talk,* edited by Deborah Tannen, 71–93. Washington, DC: Georgetown University Press.

Schegloff, Emanuel A. 1989. "Reflections on Language, Development, and the Interactional Character of Talk-in-Interaction." In *Interaction in Human Development,* edited by Marc H. Bornstein and Jerome S. Bruner, 139–153. Hillsdale, NJ: Lawrence Erlbaum.

Schegloff, Emanuel A. 1992d. "Repair After Next Turn: The Last Structurally Provided Defense of Intersubjectivity in Conversation." *American Journal of Sociology,* 97(5), 1295–1345.

Schegloff, Emanuel A. 1997. "Practices and Actions: Boundary Cases of Other-Initiated Repair." *Discourse Processes* 23 (3): 499–545.

Schegloff, Emanuel A. 2007. *Sequence Organization in Interaction: A Primer in Conversation Analysis,* Volume 1. Cambridge: Cambridge University Press.

Schegloff, Emanuel A. 2010. "Some Other 'Uh(m)'s." *Discourse Processes* 47: 130–174.

Schegloff, Emanuel A., Gail Jefferson, and Harvey Sacks. 1977. "The Preference for Self-Correction in the Organization of Repair in Conversation." *Language* 53 (2): 361–382.

Schegloff, Emanuel A., Elinor Ochs, and Sandra A. Thompson. 1996. "Introduction." In *Interaction and Grammar,* edited by Schegloff, Ochs, and Thompson. Cambridge: Cambridge University Press.

Schiffrin, Deborah. 1988. *Discourse Markers*. Cambridge: Cambridge University Press.

Searle, John R. 1990. "Collective Intentions and Actions." In *Intentions in Communications*, edited by P. Cohen, J. Morgan, and M. E. Pollack, 401–415. Cambridge: MIT Press.

Sidnell, Jack. 2010. *Conversation Analysis: An Introduction*. London: Wiley-Blackwell.

Sidnell, Jack, and Tanya Stivers, eds. 2012. *The Handbook of Conversation Analysis*. Oxford: Wiley-Blackwell.

Sperber, Dan, and Dierdre Wilson. 1995. *Relevance: Communication and Cognition*, 2nd ed. Oxford: Blackwell.

Steffenson, Sune V., and Alwin Fill. 2013. "Ecolinguistics: The State of the Art and Future Horizons." *Language Sciences* 41: 6–25.

Stivers, Tanya. 2010. "An Overview of the Question-Response System in American English Conversation." *Journal of Pragmatics* 42: 2772–2781. doi:10.1016/j.pragma.2010.04.011.

Stivers, Tanya, N. J. Enfield, Penelope Brown, Christina Englert, Makoto Hayashi, Trine Heinemann, Gertie Hoymann, et al. 2009. "Universals and Cultural Variation in Turn-Taking in Conversation." *Proceedings of the National Academy of Sciences of the United States of America* 106 (26): 10587–10592. doi:10.1073/pnas.0903616106.

Stivers, Tanya, and Jeffrey D. Robinson. 2006. "A Preference for Progressivity in Interaction." *Language in Society* 35 (3): 367–392.

Stivers, Tanya, and Federico Rossano. 2010. "Mobilizing Response." *Research on Language and Social Interaction* 43: 3–31.

Svennevig, Jan. 2008. "Trying the Easiest Solution First in Other-Initiated Repair." *Journal of Pragmatics* 40 (2): 333–348.

Takahashi, D. Y., D. Z Narayanan, and A. A. Ghazanfar. 2013. "Coupled Oscillator Dynamics of Vocal Turn-Taking Monkey." *Current Biology* 23: 2162–2168.

Tannen, Deborah. 1984. *Conversational Style: Analyzing Talk Among Friends*. Oxford: Oxford University Press.

Thompson, S. A. 1998. "A Discourse Explanation for the Cross-Linguistic Differences in the Grammar of Interrogation and Negation." In *Case, Typology and Grammar: In Honour of Barry J. Blake*, edited by Anna Siewierska and Jae Jung Song, 309–341. Amsterdam: John Benjamins.

Thompson, Sandra A., Barbara A. Fox, and Elizabeth Couper-Kuhlen. 2015. *Grammar in Everyday Talk: Building Responsive Actions*. Cambridge: Cambridge University Press.

Tomasello, Michael. 2008. *Origins of Human Communication*. Cambridge: MIT Press.

Tomasello, Michael. 2016. *A Natural History of Human Morality*. Cambridge: Harvard University Press.

Turner, Lynn H., and Richard West. 2010. *Communication Accommodation Theory: Analysis and Application*. 4th ed. New York: McGraw-Hill.

Tyack, Peter L. 2003. "Dolphins Communicate About Individual-Specific Social Relationships." In *Animal Social Complexity: Intelligence, Culture, and Individualized Societies*, edited by Frans B. M. de Waal and Peter L. Tyack, 342–361. Cambridge: Harvard University Press.

Tylor, E. B. 1889. "On a Method of Investigating the Development of Institutions; Applied to Laws of Marriage and Descent." *Journal of the Anthropological Institute of Great Britain and Ireland* 18: 245–272.

Uryu, Michiko, Sune V. Steffenson, and Claire Kramsch. 2014. "The Ecology of Intercultural Interaction: Timescales,

Temporal Ranges and Identity Dynamics." *Language Sciences* 41: 41–59. doi:10.1016/j.langsci.2013.08.006.

Ward, Nigel. 2006. "Non-Lexical Conversational Sounds in American English." *Pragmatics and Cognition* 14 (1): 129–182.

Wierzbicka, Anna. 1996. *Semantics: Primes and Universals*. Oxford: Oxford University Press.

Wierzbicka, Anna. 2003. *Cross-Cultural Pragmatics: The Semantics of Human Interaction*. Berlin: Walter de Gruyter.

Zeitlyn, David. 1995. "Divination as Dialogue: Negotiation of Meaning with Random Responses." In *Social Intelligence and Interaction: Expressions and Implications of the Social Bias in Human Intelligence*, edited by Esther N. Goody, 189–205. Cambridge: Cambridge University Press.

INDEX

University of Sydney

N. J. Enfield is Professor of Linguistics at the University of Sydney, and Director of the Sydney Social Sciences and Humanities Advanced Research Centre (SSSHARC). His work on language and the social mind has appeared in outlets like the *New York Times*, *Scientific American*, *Science*, and NPR. He has published 17 books and more than 150 scholarly articles and reviews. He lives in Sydney, Australia.